EQUITY FOR WOMEN IN SCIENCE

EQUITY FOR WOMEN IN SCIENCE

Dismantling Systemic Barriers to Advancement

CASSIDY R. SUGIMOTO

AND

VINCENT LARIVIÈRE

HARVARD UNIVERSITY PRESS

Cambridge, Massachusetts & London, England

2023

First printing

Library of Congress Cataloging-in-Publication Data

Names: Sugimoto, Cassidy R., author. | Larivière, Vincent, author.
Title: Equity for women in science : dismantling systemic barriers to
advancement / Cassidy R. Sugimoto and Vincent Larivière.
Description: Cambridge, Massachusetts : Harvard University Press, 2023. |
Includes bibliographical references and index.
Identifiers: LCCN 2022037279 | ISBN 9780674919297 (hardcover)
Subjects: LCSH: Women in science. | Sex discrimination against women. |
Sex discrimination in science. | Sex discrimination in employment.
Classification: LCC Q130 .S84 2023 | DDC 500.82—dc23/eng20221103
LC record available at https://lccn.loc.gov/2022037279

À Jean et Louise

CONTENTS

Nevertheless, she persisted.

EQUITY FOR WOMEN IN SCIENCE

Introduction

*M*y *mother is a scientist. My daughter is a scientist. I have women scientists in my lab. I publish with women. There are many women in my department.* These are the proclamations we hear as we present on gender disparities in science. These assurances are sometimes used to celebrate the increasing presence of women in the scientific workforce. Yet they also serve as a gentle protest against research on gender disparities in science. The statements suggest that disparities are a thing of the past. Optimistically, they suggest that science is now an inclusive space and, perhaps, that the lingering disparities observed may be disparities earned, inherent, or desired.

We have seen great advances in the inclusion of women in the scientific world, as in other spheres of society. Science does not occur in a cultural vacuum; it is deeply situated in the sociopolitical contexts in which it is conducted. It is no surprise, therefore, that the growing recognition of women in science occurred in parallel with their acknowledgment in the political realm. In 1893, New Zealand became the first self-governing country to grant women the vote. It took a few more decades until the same rights were granted to women in Britain (1918) and in the United States (1920). Scientific and educational societies followed the zeitgeist: in 1919, the Geological Society in the United Kingdom voted to admit women; the UK Chemical Society admitted the first woman in 1920; and Cambridge awarded its first degree to a woman in 1921.[1] Not all institutions responded as swiftly. The Royal Society of London did not elect women to full membership until 1945—the same year that Harvard Medical School began admitting women—and the Paris Academy of Science did not admit its first woman until 1979.[2] Progress has been slow and uneven across the globe, but the steady steps toward gender parity are marked by these milestones.[3]

Parity represents the degree to which women are equally represented in social institutions. Parity, however, is not always associated with equity. For example, Nordic societies are often heralded as paragons of parity: in the World Economic Forum 2021 report, Iceland, Finland, and Norway were ranked first through third and Sweden was ranked fifth in terms of gender parity.[4] Yet these countries have a disproportionately high rate of intimate partner violence against women; much higher than the Organisation for Economic Cooperation and Development average.[5] The phenomenon was identified as the "Nordic paradox": How can a society with such high levels of economic parity demonstrate such atrocious inequities in the social realm? Many will argue that the rankings themselves are flawed. This is, of course, part of the story. However, it is also true that merely introducing women—to the labor market, politics, or science—does not itself remove bias and discrimination, in public or private spheres. As the #MeToo movement has shown, even in countries that have exhibited the highest growth in the educational and scientific achievement of women, institutions have not fostered cultures of inclusivity. This reinforces sociologist Mary Frank Fox's unambiguous argument: "Increasing the number of women will not necessarily change patterns of gender and hierarchy in science."[6] While achieving parity in participation is an important step, it does not mean that equity has been reached. Furthermore, parity may not come from a place of progress, but rather of displacement and devaluation. As classicist Mary Beard notes in *Women and Power,*

> There are plenty of league tables charting the proportion of women within national legislatures. At the very top comes Rwanda, where more than 60 per cent of the members of the legislature are women, while the UK is almost fifty places down, at roughly 30 per cent. Strikingly, the Saudi Arabian National Council has a higher proportion of women than the US Congress. It is hard not to lament some of these figures and applaud others, and a lot has rightly been made of the role of women in post-civil war Rwanda. But I do wonder if, in some places, the presence of large numbers of women in parliament means that parliament is where the power is *not*.[7]

The same is true in science as in parliament. Parity does not necessarily equate with scientific capital. As we will demonstrate in this book, countries that are closer to gender parity in scientific production tend to be those that experienced high degrees of brain drain—where the men have left the country for greater scientific resources. Disciplines with parity (or those dominated by women) tend to be lower cited and relegated to the bottom tiers of the academic hierarchy. This is not a novel finding: occupational studies have long shown that feminized professions— that is, those with high degrees of women (such as social work, education, and librarianship)—tend to be depressed in terms of both salary and social capital.[8] This disparity is not merely an artifact of the types of work that may be done by men relative to women: when women enter fields in greater numbers, pay declines.[9] The corollary is also true: when salaries in occupations decline, the proportion of women increases. The difference is perhaps most striking when examining jobs of similar education and skill levels, differentiated only by gender. For example, janitors (predominantly men) earn 22% more than maids and housecleaners—who are primarily women.[10] We seek to examine this phenomenon within the realm of scientific production—to understand, from a quantitative perspective, degrees of parity across disciplines and countries, how this relates to scientific labor, and the factors that may underlie disparities observed.

The ethos of science states, as per sociologist Robert K. Merton, that science should be open to all; that nothing other than the lack of skills or knowledge should prevent people from participating in scientific activities.[11] It is with this principle that many organizations have worked diligently toward increasing the participation of women and other minorities to match their representation in society. Contemporary examples of parity, however, tend not to be laudable exemplars, but rather demonstrations of devaluation. We refer to this phenomenon as the *parity paradox:* wherein striving toward parity does not result in equity. This paradox presents a strong policy dilemma—parity without corresponding equity devalues the labor it seeks to reward.

Scientific labor has traditionally been defined by men and reinforced the contributions they made. Science is an inherently hierarchical space, and, according to Fox, "gendered relationships are hierarchical inasmuch as women and men are not simply social groups neutrally distinguished from each other, but rather, are differentially ranked and evaluated according to a standard of masculine norms and behavior." This is not to suggest that women have not historically contributed to science; they were active in research long before they were recognized by scientific institutions. As Fox has remarked, "Women have long been 'in science,' but not central to science."[12] In some ways, their participation was tolerated more judiciously before the professionalization of science.[13] However, as science sought to establish its credibility among other professions in the early twentieth century, the presence of women and minorities threatened that professionalization.[14]

Occupational terminology can be quite revealing. In the early twentieth century, the two competing terms used to refer to a scientist were *man of science* and *scientific worker*. These terms were meant to distinguish professional scientists from the amateurs who preceded them; to mark with distinction those who were qualified to work in science. Distinguished scientists would then be listed in reference works, such as the *American Men of Science,* which chronicled scientists in North America. As the number of women employed in science began to grow, the exclusionary and imprecise terminology came under fire. A 1924 letter to the editor in the generalist journal *Nature* bemoaned the term *men of science* and called for *Nature* to adopt the more precise and inclusive term *scientist*. The term *scientist* had been coined nearly one hundred years prior by William Whewell, in his review of an astronomy article by Mary Somerville—the first woman member of the Royal Astronomical Society.[15] It was not until 1971, however, that the *American Men of Science* was retitled as the *American Men and Women of Science. Nature* did not respond until the next century: the journal adopted a new mission statement removing the phrase "men of science" in 2000.[16] This exemplifies the deeply rooted and hierarchical structures of power relations in science.

This book is an examination of the gendered nature of scientific production, labor, and reward. We seek to describe the disparities that exist and reveal some of the mechanisms underlying gender disparities and corresponding inequities in science. We deconstruct the parity paradox by examining the persistence of women in science across time and place and exploring a deeper and more contextualized understanding of disparity in scientific labor. It is critical to both understand the contemporary role of women in science and to be able to identify barriers to success. Only then can we move toward a scientific ecosystem in which women are both included and valued.

Motivation

In 1974, Ruth Hubbard was the first woman biology professor to be awarded tenure at Harvard.[17] In an interview with the *New York Times* following her tenure, she noted, "Women and nonwhite, working-class and poor men have largely been outside the process of science-making. Though we have been described by scientists, by and large we have not been the describers and definers of scientific reality. We have not formulated the questions scientists ask, nor have we answered them. This undoubtedly has affected the content of science, but it has also affected the social context and the ambience in which science is done."[18]

It matters who is making science. In this book, we demonstrate that women are underrepresented in almost every field of science. When they are present, they are often relegated to the periphery or to technical rather than conceptual roles. Our research supports Hubbard's claims: women have historically neither asked nor answered the questions of science. But does this matter? A social justice perspective would argue that this underrepresentation is problematic in that it creates barriers where certain populations do not have access to the full range of occupations. However, does this fundamentally alter the content of science and what Hubbard termed the "scientific reality"? Does it matter who is asking the questions? Does this change what we know about ourselves and the world around us?

These were the motivating questions for a study we conducted on sex and gender in biomedical research.[19] Our study sought to analyze whether the inclusion of women in biomedical research affects the populations that were studied, focusing specifically on sex. A large body of research has demonstrated sex differences at the genetic, cellular, biomedical, and physiological levels.[20] Despite this, there have been disparities in the inclusion of sex as an analytic variable in biomedical research. We found some areas of improvement: females, who were historically underrepresented in large-scale clinical trials, are now included at greater rates, and sex reporting is improving in biomedical sciences.[21] Although increasing, the rate of sex reporting for preclinical studies remains low. However, our results demonstrated that when a study was women led, it was much more likely to report on sex and to include female samples or populations in the study. This suggests that diversity in the scientific workforce is essential to produce the most rigorous and effective medical research: when we have women in biomedicine, we are more likely to have biomedical research that looks specifically at females. Quoting science journalist Angela Saini in *Inferior,* a popular account of research on women, "Having more women in science is already changing how science is done. Questions are being asked that were never asked before. Assumptions are being challenged. Old ideas are giving way to new ones."[22]

One may contend that these disparities are historical artifacts that will naturally change over time. There is, indeed, a gradual move toward parity. Over the last decade, the proportion of women authorships increased in every discipline, albeit with different growth rates (Figure I.1). From these rates, we can calculate a rough estimate of the time it will take to reach parity. As we were writing this book, psychology reached parity (50%), increasing from 43% in 2008. At this rate, parity will be reached in the social sciences and in arts and humanities in 2044, and clinical medicine in 2049. Earth and space sciences would be the first field in the natural sciences to reach parity (2063), followed by biology (2069), biomedical research (2074), and chemistry (2087). Other disciplines of the natural sciences would still need, at the current growth rate, a century or more to reach parity: 2144 for engineering, 2146 for mathematics, and 2158 for physics.[23] This is not a promising story. Furthermore, increases in parity do not necessarily account for the hierarchical power structures in science, which mediate question formulation and investigation.[24]

These data suggest that without strong interventions, several generations will pass before men and women have equal opportunities to shape scientific knowledge. The implications are obvious in health but are equally important in other areas. Engineering, for example, has one of the lowest rates of women, and this is not without consequences. Car manufacturing and testing is one example of the consequences of gender domination in a field. Women are 47% more likely to be seriously injured and 17% more likely to die in a car crash. This has been attributed, at least in part, to the design and testing of automobiles. Female-typed crash dummies were not introduced until 2003 and were cast as five-foot-tall, 110-pound scaled-down male test dummies. They were then only tested in the passenger seat.[25] Women who are driving or outside these dimensions (such as those who are pregnant) were not considered in the engineering of the vehicle. In fact, the manual manipulations that women drivers make—sitting more upright and closer to the dashboard—are seen as acts of "noncompliance" that place them at greater risk. The construction of the vehicle suggests that women are not "fit" to be drivers. A

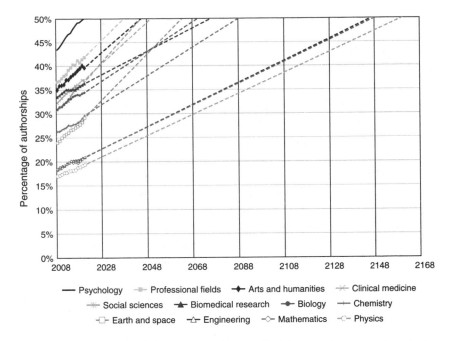

FIGURE I.1. Percentage of women authorships, by discipline, 2008–2020, projected using 2008–2020 linear growth until 2168.

similar issue developed when the first all-women spacewalk was scheduled. Within days of the scheduled flight, NASA realized it did not have appropriately sized space suits for the two women and had to replace one woman with a man.[26] The message was clear: women were not the right size to be astronauts. These stories suggest that the full inclusion of women into science will dramatically affect all sectors of society. When science fits women, women are more likely to ask the questions and make the innovations necessary to right-size all domains for women.

A Scientometric Approach

A precise measure of scientific production is necessary to provide a global and contemporary account of women's work in science. Unfortunately, generating new knowledge does not naturally lend itself to the same types of "objective" input and output indicators observed in other sectors. In a widget factory, one might count the number of hours worked and the number of widgets produced. New knowledge, however, rarely takes the form of tangible, standardized artifacts that can be counted in an industrial fashion. Despite this difficulty, the operationalization of scientific labor has largely adopted industrial metaphors: for example, we often measure the "production" of the "scientific workforce" to justify a "return on investment" for resources provided to scientists. To this end, several systematic surveys are conducted—mostly by governmental agencies and international organizations, including the United Nations Education, Scientific and Cultural Organization and the Organisation for Economic Cooperation and Development—to measure the relative strength of science in select countries. These data provide information on, inter alia, the size and composition of the scientific workforce, gathered largely from graduation and labor statistics. Along with data on R&D funding, which are also generally obtained from survey, these data are intended to serve as a proxy for input. However, while these surveys detail the production *of* scientists, they do not account for the production *by* scientists.

Measuring research *activity* directly is a surprisingly difficult task to do at scale. Besides occasional contract work, there are relatively few billable hours in science and no comprehensive analysis of the time that people contribute to scientific labor.[27] Therefore, the production of scientific works—in the form of journal articles, books, and other documents—generally serves as an indicator of the *output* of research activities.[28] Given that there are relatively few comprehensive indexes on genres of production such as books and conference papers, authorship of journal articles functions as the primary measurement of scientific labor. There are, of course, issues with this operationalization. The first is assuming that all scientific work ends in a publicly disseminated document. Many research projects fail and are placed in the "file drawer" of negative results.[29] Competitive interests may lead to the lack of diffusion of research activity with commercial applicability.[30] Furthermore, although journal articles are the modal production of most fields, only focusing on journal articles underrepresents the scientific activity of some disciplines that favor dissemination in books and conference proceedings or where commercialization activities are valorized.[31] These genres—particularly conference proceedings and patenting—tend to be more prevalent in fields dominated by men.[32] Finally, and perhaps most importantly, there is not a standard amount of time or labor that one can attribute to the production of an article: this will vary across disciplines and even within a single researcher's oeuvre due to a number of factors (such as scope of inquiry, efficiency of collaboration, or methodology). Despite these limitations, the production of journal articles remains the most efficient proxy at present for large-scale and cross-disciplinary analyses of research activity. Therefore, we rely largely on journal articles—specifically metadata as indexed in Web of Science, a large-scale bibliometric database—as a proxy for the participation of women in scientific work.

We use the term *science* and related terms in the title and throughout the pages of this book. It is critical to clarify that we do not mean this in the exclusionary way to refer to only those sciences classed as "hard" or "natural." As historian of science Derek de Solla Price observed, "The peculiar English term 'science' acts as a barrier to the belief that subjects other than physics, chemistry, and biology (in that order?) can be scientific." He continues, "In other languages, the words nauka and Wissenschaft carry a breadth of the totality of learning."[33] It is notable that the term *scientometrics* is translated from one of these root terms. In 1925, Polish sociologist Florian Znaniecki coined the term *naukoznawtswo,* which translates as "science studies" or, more precisely, "science connoisseurship." Subsequent Polish scholars used the phrase *nauka or nauce* to refer to the "science of science." In 1966, Russian philosopher Vasiliy Vasilevich Nalimov coined *naukometriya* and later defined it as "the information process which uses quantitative methods for the exploration of science."[34] Nalimov credits John Desmond Bernal as the founding father of this field; de Solla Price also argues that Bernal's *Social Function of Science* was the founding primer for the "science of science." This suggests that the two terms—*scientometrics* and *science of science*—developed without clear distinction and can be considered synonymous. Quoting from de Solla Price, "This new study might be called 'history, philosophy, sociology, psychology, economics, political science and operations research (etc.) of science, technology, medicine (etc.).' We prefer to dub it 'Science of Science,' for then the repeated word serves as a constant reminder that science must run the entire gamut of its meanings in both contexts."[35] We employ the fullness of that spectrum in our use of the term science.

Bibliometric data are included in a few reports alongside workforce and funding information, for example, in the Science and Engineering Indicators report produced by the National Science Foundation. This allows for rough estimations of "return on investment"—that is, the relationship between the amount of investment in research and the associated output. In the 2020 Science and Engineering Indicators report, data were analyzed by gender as well, detailing the proportion of women authorship for science and engineering publications for a sample of countries. Data such as these provide snapshots of global gender disparities in scientific production but often suffer from several weaknesses, most notably datedness, incompleteness, and selection bias. The time that it takes to compile and analyze these data invariably makes them dated upon arrival; furthermore, these data tend to purposively showcase certain high-performing or peer countries and fail to provide comprehensive global analyses.

A scientometric approach provides certain advantages for global and contemporary insights on scientific production. By standardizing scientometric metadata, it is possible to perform rich analyses that take into account differences across countries and disciplines. To study gender, the most important metadata is that of authorship. The relative presence of women on the bylines of articles is, in many ways, an even more important indicator than the number of women in the scientific workforce. This measurement provides evidence of the overall contribution of women to scientific literature and how women's labor is made manifest to the scientific community. Authorship demonstrates who has a voice in science—who gets to participate in the labor of science and who is acknowledged—and reaps the rewards of the results of that work. In an ideal scenario, authorship signals to the scientific community that an individual is associated with the scientific labor underlying that document.

Authorship is a gateway into the contemporary reward system of science and therefore, a focal point of the present analyses. The names on the byline influence how an article is reviewed and received by the scientific community. It is also critical for the reward system of science: scholars are not assessed by their hours in the lab or their search through the archives, but by the *outputs* of this labor. Authorship is, therefore, the coin of the realm, providing the currency for the economy of academic reputation.[36] The academic market is built on this concept of capital, wherein one gains authorship for contributing to scientific work and is rewarded for that contribution when it generates citations. This makes authorship a critical lens through which we can examine parity and equity in scholarship.

Two recent methodological contributions have made this book possible. First, advances in gender assignment techniques—that is, estimating the gender of author based on their given names and, in the case of some countries, family names—allowed gender-based analysis of scholarly production at scale, with sufficient precision and recall (see appendix).[37] Second, advances in author disambiguation techniques—that is, the attribution of a body of work to a specific researcher, based on its characteristics (affiliations, disciplines, cited references, or coauthors)—have provided opportunities for examining individual researchers' career trajectories and are used in this book to assess research productivity and mobility.[38] These contributions have made it possible to move from institution-level, national, and disciplinary analyses to large-scale international and multidisciplinary analyses of gender disparities in science, and to move from article-level to individual-level analyses.

The scientometric approach, however, is not without limitations. The published scientific document stands as a proxy for a host of complex interpersonal and cognition processes of knowledge production that precede it. The document can provide insights into these processes, but there are several issues on which it is silent. For example, despite the rise of several mechanisms for ensuring ethical authorship practices, the document cannot tell us whether there was justice in the allocation of authorship. For this, we need to ask the authors.[39] Moreover, the many caveats to the Web of Science as a bibliometric data source apply. Despite its high-quality metadata, the Web of Science has a weak coverage of journals published in languages other than English and from non-English-speaking countries, and given its focus on journal articles, it has a poor coverage of disciplines from the social sciences and humanities as well as of computer science and engineering, which disseminate knowledge through books, book chapters, and conference proceedings.[40] Furthermore, algorithmic assignment of gender reinforces the binary characterization of gender and may introduce inaccuracies at the individual level. We acknowledge that gender identities are expansive, and that this distinction is inherently problematic. Therefore, we have complemented our bibliometric analysis with surveys to both extend and validate our approach. Furthermore, we stress that our goal is not to determine the gender of any individual author but to provide a rough estimation at the macro level to begin to unravel gendered distinctions in science. We hope that this work paves the way for future studies that can examine the plurality of gender and other identities.

We acknowledge the volumes of work in history, sociology, and related fields that have sought to provide a deeper understanding of the role of women in science. This book is deeply indebted to the work of several scholars who have sought to chronicle the lives of women in science across the ages.[41] As the first comprehensive scientometric account of women in science, this book should be read not as a replacement for previous work but as a complement to it. Our account of women in science advances the conversation by providing a large-scale description of the role of women in contemporary science. By examining several factors—such as collaboration, mobility, and funding—we can provide both a diagnostic of the degree of disparity for women in science and a policy-relevant account of the barriers to participation in the scientific workforce. Such information is vital for the present and future scientific community—those pursuing scientific careers, mentoring scientists, and all who serve as gate-keepers to the production of science.

Organization of the Book

Each chapter begins with a short set of anecdotes and profiles of women in science. These select exemplars provide historical context to guide the reading of the empirical analyses.[42] These selections were made purposively, to find examples that tied the theme of the chapter together across time and place. Documentation of women in science tends to emphasize white, Western women from historically dominant scientific disciplines (for example, physics, astronomy, and chemistry). It is our hope that the profiles in this book make incremental improvements toward increasing the visibility of a more diverse range of women. We acknowledge the importance of intersectionality: race, era, country, and discipline all have dramatic implications for the experience of women in science. Where relevant, we describe contextual factors of specific countries or disciplines that serve as exemplars in understanding the complex sociocultural factors that contribute to gender disparities in scientific production. These accounts are not meant to be comprehensive histories of science making in these countries; rather, their function is to highlight the rich geopolitical landscapes that must be considered when interpreting scientometric data. Intersectionality and the geopolitical contexts of science are all essential components of understanding the parity paradox and moving toward a more equitable science system. One limitation of our analysis is that we focus on one dimension of social identity—gender—and omit other attributes, such as race and ethnicity. This is due to our emphasis on the position of women in science across the globe. Studies of race, for example, must be conducted within the frame of a single country or region, where racial constructs are shared. We provide an example of such an intersectional analysis in Chapter 8.

The core contribution of the book is the contemporary scientometric analysis of gender. We focus on six main facets of research that can be studied using bibliometric metadata: production, collaboration, contributorship, funding, mobility, and impact. Chapter 1 (Production) sets the scene by providing a general diagnostic on the place of women on the byline of scholarly articles, looking at both production—the proportion of scholarship that is produced by women, in the aggregate, and productivity—and how much each individual woman produces. Differences across countries and disciplines are disentangled, with a focus on country-level variables as potential correlates with gender disparities. This chapter serves as the basis for the rest of the book, as authorship is the key anchor for studying women through a scientometric lens. Chapter 2 (Collaboration) builds on these results by looking at multiple authorship, also referred to as collaboration. Team organization is essential for understanding scientific trajectories. The chapter examines the gendered nature of authorship ordering in collaborative teams and the interplay between gender, leadership, and team size. Furthermore, this chapter begins to unravel the relationship between authorship and labor contribution, explored in further detail in the chapter that follows. Chapter 3 (Contributorship) explores how men and women within a scientific team are associated with different labor roles and the effect of task specialization on careers.

Using both bibliometric data and sources from funding agencies, Chapter 4 (Funding) examines gender differences in grant support for research, focusing on gendered differences in the percentage of funded research projects as well as success rates and funding amounts. Funding is increasingly used as an output indicator, rather than an input indicator, with strong implications for retention in science and scientific success. Chapter 5 (Mobility) moves to a relatively new area of research in bibliometric studies—scientific mobility—and uses affiliation data on papers to examine gendered differences in affiliation changes across countries and the relationship with other variables, such as production and impact. Chapter 6 (Scientific Impact) examines the citation received for men's and women's papers and how this is mediated by the factors examined in other chapters, including collaboration. Self-citation is explored as one potential explanation for the gender gap observed. Chapter 7 (Social Institutions) examines the social and institutional structures that contribute to gender disparities in science. Chapter 8 (Recommendations and Conclusions) provides a series of recommendations for various stakeholders, including researchers, institutions, research funders, publishers, and journalists. Taken together, this book provides an overview of the state of equity for women in science and mechanisms that may serve to dismantle systemic barriers to advancement.

Chapter 1

Production

In 2010, the Royal Society published a list of ten British women who had the greatest influence on science.[1] This list contains contemporary scientists such as geneticist Anne McLaren and Nobel laureate biochemist Dorothy Hodgkin. Most of these women have high educational achievements and laudations to accompany their robust pedigrees. One woman, however, is distinctively unadorned with academic accolades.

Mary Anning was born in Great Britain shortly before the turn of the nineteenth century (1799–1847). Her childhood and adolescence did not involve tutors or schooling, but rather days exploring in Lyme Regis on the southern shore of Britain. Her father was an occasional fossil collector and taught his daughter the skills of collecting and dealing. This provided a modest income for the family, as they were lucky enough to live near a rich Jurassic-era marine fossil bed. Around age twelve, Anning found the first full ichthyosaurus to be known to the scientific community. She subsequently discovered the first plesiosaur and pterodactyl fossils. These fossils were purchased by museums and collectors, many of whom noted her strong knowledge of paleontology. Despite this, she received little recognition for her work, besides an obituary notice from the director of Her Majesty's Geological Survey and president of the Geological Society of London—an association to which neither she nor another woman was admitted until 1904. Her sole published work is a letter to the editor in the *Magazine of Natural History*. Despite having her fossils featured in the Natural History Museum and other prestigious institutions, her contributions to paleontology were largely forgotten after her death. Her gender and social class prevented her from being able to contribute to the permanent scientific record. Anning was engaged in research "activity," but she did not "produce" science.

It took another generation for women's work to be made visible in scientific publications. British engineer, mathematician, and physicist Hertha Ayrton (1854–1923) was also featured on the list of notable British scientists. Among her scientific honors is the distinction of being the first woman to read her own research before the Royal Society. Unlike Anning, Ayrton was well schooled, attending Cambridge University and the University of London. As with many successful scholars of her era, Ayrton's entrance into the scientific community was facilitated through marriage: she met and married her professor, William Ayrton, when she was attending classes at Finsbury Technical College in the late 1880s. The marriage, however, hampered some of her subsequent success. Although she was nominated as a fellow of the Royal Society of London, she could not be elected due to her marital status. Her marriage also led to some disquieting authorship issues. Ayrton's aging husband was unable to maintain his productivity; to facilitate, Ayrton continued doing his work and publishing under his name in parallel with her own work. She therefore stands at an interesting intersection in the history of scientific production for women: her marriage provided her entrance into science yet also made many of her contributions invisible. Despite these constraints, she eventually developed an independent voice in the scientific community, receiving several notable awards and publishing in the most reputable journals of the day.

One of the leading publication venues then (and now) was *Nature,* a generalist scientific magazine founded in 1869.[2] In Ayrton's time, the editor in chief (and founder of the journal) was Sir Joseph Norman Lockyer, who was married to a strong advocate for women's rights and a noted suffragette, Mary Broadhurst Lockyer. When Ayrton was considered for the Royal Society, Mary Lockyer strongly advocated in her favor. Her husband was similarly supportive of women's entrance into these societies. In 1904 and 1908, women chemists petitioned the Royal Chemical Society to admit women fellows.[3] A *Nature* editorial ran in support of the 1908 petition, noting the positive contribution of women scientists to the journal: "It cannot be denied that women have contributed their fair share of original communications. Indeed, in proportion to their numbers they have shown themselves to be among the most active and successful of investigators. The society consents to publish their work, which redounds to its credit."[4]

Nature editorials also expressed support for women scientists seeking admittance to other societies, such as the Geological Society, where Anning might have been admitted were she of another century. Not all editors were equally supportive. The editor who succeeded Lockyer at *Nature,* Richard Gregory, was far less sympathetic to the cause. He published an obituary of Ayrton by fellow chemist Henry E. Armstrong strongly suggesting that Ayrton's success should be credited to her husband: "I never saw reason to believe that she was original in any special degree; indeed, I always thought that she was far more subject to her husband's lead than either he or she imagined."[5]

Progress toward equity was slow. In 1924, a letter to the editor from renowned physicist Norman R. Campbell requested that *Nature* replace the offensive term *man of science* with the more inclusive term *scientist.*[6] Gregory responded that he would inquire among many "distinguished men of science" on their opinions on the matter. As noted in the introduction, it was not until 2000 that *Nature* adopted a new mission statement, eliminating the reference to "men of science." It took nearly two more decades after this change for a woman to be appointed to the helm: *Nature* hired its first woman editor in chief, geneticist Magdalena Skipper, in 2018. More than two centuries passed, from Anning to Skipper, for women not only to be recognized for producing science but to serve in the highest echelons of gatekeeping.

Women and the Professionalization of Disciplines

The stories of Anning and Ayrton and the controversy in the pages of *Nature* foreshadow tension between the professionalization and feminization of scientific disciplines. Before the twentieth century, science was associated with amateurism, which allowed wealthy women and those who married scientists to participate in scientific activities. Beginning in the 1870s, women increased their membership in scientific organizations and began obtaining employment in museums and observatories.[7] By the end of the nineteenth century, however, science began a process of professionalization that served to decrease women's access to scholarship.[8] When science became codified as a professional— and therefore masculine—domain, women were further isolated from participation.

The rise of higher education and the expansion of employment for women happened in parallel with this professionalization. Several institutions across the world opened their doors to women in the late nineteenth century, and by 1910, a large share of universities allowed women to receive degrees across the disciplinary spectrum. This led to an environment where women were being educated but not employed or advanced at the same rate as men. Women were therefore corralled into careers that were perceived as appropriate for their "special talents." Gender segregation in labor, well established in other professional sectors, intensified in science. Women initially took their science degrees into museums, botanical gardens, and other cultural institutions where they could receive modest employment. Over time, the rise of larger, collaborative teams opened new opportunities for women as science assistants.[9] For example, women's roles as "computers" in astronomy and physics confirmed their ability to conduct patient and painstaking labor in the lab.

As women matriculated at higher rates, they were more likely to enter disciplines associated with care and domesticity—such as education, child psychology, librarianship, social work, and the newly emerging field of home economics.[10] It is not surprising that Nobel laureate Marie Curie was cast in a caregiving role in her fund-raising trips in North America: rather than being touted for her scientific discoveries, she was aligned with the potential to cure cancer and her work on the military front. It was easier to place a woman in a care-related role than to perceive her as a scientist.[11]

Caregiving fields were feminized at the start of the twentieth century and remain so. In contemporary research, women account for less than a third of authorships (30.9%), with few research specialties where women are in the majority.[12] In practical terms, this means that for every woman scientist, there are about 2.3 men. As our data show, only 10 of the 143 specialties in the National Science Foundation journal classification have more authorships by women than men (Figure 1.1): nursing (75.8%), social work (64.1%), speech-language pathology and audiology (62.2%), developmental and child psychology (61.2%), public health (52.2%), rehabilitation (53.7%), social sciences, biomedicine (53.6%), education (53.5%), geriatrics and gerontology (53.5%), and nutrition and dietetics (50.3%).[13] All these woman-dominant fields have a distinctive service aspect to them, thereby reinforcing the perception of women as more suitable for care-oriented disciplines.[14]

The disciplines and specialties with stronger representation from women are often associated with depressed salaries, leading to a cyclical reinforcement rooted in this history: only women were willing to work in fields with lower salaries, so those fields with a high proportion of women could offer lower incomes.[15] Such devaluation of women's work is common across the world. In the United States, for instance, studies have shown that occupations with a greater share of women pay less, and as more women enter a field, the pay drops.[16] The opposite is also true. Programming was historically done by women "computers," but when men outnumbered women in the field, the pay and prestige increased.[17]

On the other side of the spectrum, twenty specialties have less than 20% women authorships. Except for one—orthopedics—all are in the disciplines of engineering, mathematics, and physics: aerospace technology (13.9%), nuclear and particle physics (15.2%), mechanical engineering (15.6%), fluids and plasma physics (15.9%), and nuclear technology (16.2%).[18] Most of the disciplines with a lower percentage of women are shrouded in certain myths of the innate talent and brilliance required—that is, an inflated notion of the skill level necessary for entry and performance—and are also often associated with notions of competition and solitary work that may discourage participation from women.[19]

Similar mechanisms can be observed in the humanities: for example, the discipline with the lowest participation of women is philosophy, where associations of the field with masculinity serve as a deterrent for participation.[20] In social sciences, women are underrepresented in economics, which has a long history of hostility toward women scholars.[21] A certain proportion of women in a discipline may be required to dispel myths and prevent cultural barriers to entry. For example, a study by sociologists Catherine Riegle-Crumb and Chelsea Moore found that the dominant predictor of girls' propensity to take high school physics courses in the United States was the percentage of women in the community employed in STEM careers.[22] A similar conclusion was reached by scholars examining undergraduate majors and degree obtainment at US institutions: they found an association between the percentage of women faculty in a department and the percentage of women receiving a degree.[23] The fact that "degrees received" was a more sensitive variable than "percentage of majors" suggests the importance of role models not only for recruitment into a field but also for successful career outcomes within the field. Overall, we observe that representation matters: it is important for individuals to be

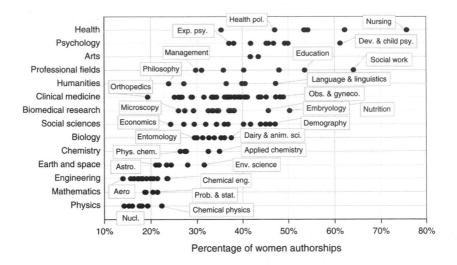

FIGURE 1.1. Percentage of women authorships, by National Science Foundation specialty and discipline, 2008–2020. Data from Web of Science core collection.

able to imagine themselves in professions through the embodiment of others who carry their same characteristics.

The Endless Productivity Puzzle

Authorship is a function not only of the number of women in the scientific system but also of their individual productivity. The distinction between the share of authorships and productivity is an important one. The percentage of authorship presents the contribution to scientific knowledge from women, *in the aggregate.* It provides the average percentage of all authors on a paper who are women but does not distinguish between a few women authoring many papers or many women each writing a single paper. Productivity, on the other hand, expresses how many papers are written by unique individuals. These distinctions provide different lenses on gendered production in science—one at the universal level and the other at the individual level. As sociologist Mary Frank Fox has observed, "Until we understand factors that are associated with productivity, and variation in productivity by gender, we can neither assess nor correct inequities in rewards, including rank, promotion, and salary . . . because publication productivity operates as both cause and effect of status in science . . . Productivity reflects women's depressed rank and status, and partially accounts for it."[24]

For all disciplines and researchers who published at least once between 2008 and 2020, we observe a sizable gap in productivity, with men publishing over the period an average of one paper more than women (4.0 versus 3.2) (Figure 1.2, left panel). For all disciplines, save arts and health, men published at least 20% more than women, with differences that reach as high as 50% more in psychology and more than a third as many articles in physics, mathematics, and social sciences. One explanation could be the gendered productivity bias that comes with age: that is, that there are more men in senior ranks and these individuals are more productive.[25] As heads of laboratories, these individuals benefit from a larger network of collaborators, which, in turn, is likely to increase publication counts.[26] To control for age differences, we examined the research productivity of a cohort of researchers who published their first paper in 2008 (Figure 1.2, right panel).[27] When only the 2008 cohort is analyzed, the productivity gender gap is much smaller, with men publishing 4.2 papers and women 3.9 over the 2008–2020 period (all disciplines combined). The arts exhibit slightly higher productivity for women than for men, and the gender gap is less than 10% in clinical medicine (3.6%), biomedical research (8.3%), biology (8.9%), and earth and space (9.9%) when the cohort of younger researchers is considered. This is likely an effect of the influx of women into these disciplines: As women have become more prominent

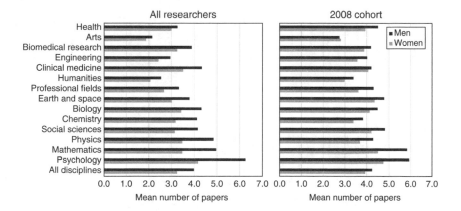

FIGURE 1.2. Total number of papers published by men and women who published their first paper in 2008 (right panel) and for all researchers who have published at least one paper over the 2008–2020 period (left panel), by discipline. Data from Web of Science core collection.

in disciplines related to medicine, average productivity for women has increased. This may reinforce the circular benefits of parity: when there are more women in a field, the climate for women improves and provides opportunities for them to thrive.

Given the wide variations in rank structures across countries and lack of standardized data, it is impossible to control for rank in these analyses. Controlling for year of first publication can be useful in creating a proxy for seniority, but not for rank, given that women are less likely to attain the rank of full professor (in the United States) or to achieve it as

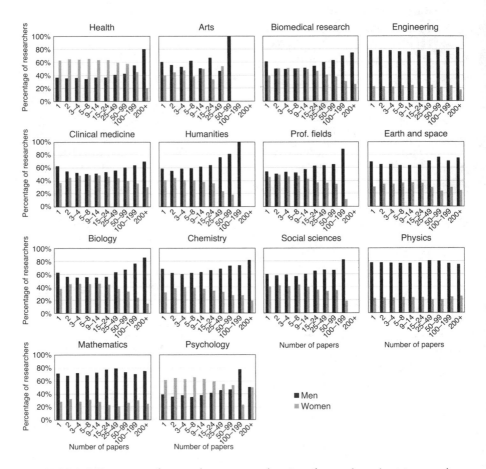

FIGURE 1.3. Percentage of men and women, as a function of research productivity over the 2008–2020 period, for the cohort of researchers who published their first paper in 2008, by discipline. Data from Web of Science core collection.

quickly.[28] Despite these limitations, our results provide some evidence to support previous findings of the mediating effect of rank on gender and productivity, suggesting that interventions focused on senior faculty would have strong effects on the field.[29]

Averages, of course, obscure differences in the distribution of research productivity. One way to examine this is to measure, by field, the percentage of women across different levels of individual research productivity, from those who have only authored a single paper to women with more than 200 papers over the 2008–2020 period (Figure 1.3). Some fields are relatively static in the proportion of women by research productivity: there is little difference in engineering and physics, for example, in the relative share of women who have one paper and those with more than 200. However, there are more women than men in all categories of production in health and psychology until the number of papers reaches 50—at this point, there are proportionally more men than women. The life sciences fields (biomedical research, clinical medicine, and biology) demonstrate inverted U-shaped curves, where there is a greater disparity at the low and high ends of production but closer to parity in the middle range of production. Given that these data are taken from a single cohort (those who published their first paper in 2008), one cannot interpret the results as merely a slow progression through time, with the influx of women changing the distribution of some fields. Rather, this suggests different patterns of production at the intersection of gender and discipline. The mechanisms that cause this continued "productivity puzzle" warrant deeper evaluation, particularly as they pertain to expectations for rank advancement in academia.[30]

Women in the World

Rates of women authorship vary by country, with few countries reaching equal representation by gender. To visualize how countries differ in their percentage of women authorships, we present two world maps: one that demonstrates how countries diverge from absolute gender parity (Figure 1.4a), and the other showing how countries diverge from the global mean percentage of women authorships of 30.9% (Figure 1.4b). Five countries—the United States, China, Japan, Germany, and the United Kingdom—account for more than half of all fractionalized authorships, thereby strongly influencing the world average (30.9%). Among these high-producing countries, Japan has the lowest rate of women authorship (17%), falling far below the world average and lower than other Asian countries. More than a quarter of Chinese authorships are women (26%), placing China at a nearly identical gender rate to Germany's (27%).[31] The United States stands above the world average, with 33% women authorship—a ratio of two men for each woman in the research system— while the United Kingdom's percentage of women authorship is slightly higher than the world average at 32%.

Of all countries and territories, only ten have an equal or higher percentage of women authorships, though eight of these countries are relatively small, with fewer than 200 fractionalized authorships over the thirteen-year period studied. The two countries with more than 200 fractionalized authorships and 50% or more women authorships are Latvia (51%) and Romania (50%). Coming close to parity are North Macedonia, Argentina, Bulgaria, Ukraine, Croatia, and Serbia, which obtain above 48% women authorships. The unique status of former Yugoslavian countries is particularly notable. In 1939, before Josip Broz Tito's communist regime, only 19% of students enrolled in institutes of higher education were women. With a commitment to promoting gender equality, the new communist regime opened higher education to women, thus increasing their percentage to nearly 30% by 1961–1962, and then to more than 40% by 1973–1974.[32] Similar trends are also observed for science-related occupations: in the 1970s and 1980s, women accounted for 43%–44% of the scientific workforce—sizably above the Yugoslav average for all sectors of the economy combined (35.5% in 1980), and above the average participation of women in most countries.[33] This is in keeping with the history of several former communist countries, which have a strong history of incorporating women into the workforce, including academe.

Despite not reaching parity in scientific production, several countries in North and South America are above global rates, with Argentina (48%) and Brazil (41%) nearing parity. Most European countries are also above average, particularly Poland (44%), Portugal (44%), Spain (40%), Scandinavian countries (35%–42%), and Eastern European nations (39%– 51%). German-speaking countries—Germany (27%), Austria (28%), and Switzerland (27%)—as well as Greece (29%) and Hungary (29%), are the exception, with percentages of women authorships below the world average. These groupings reflect the cultural aspects of gender in science and suggest that religion, history, and political systems have a demonstrable effect on the composition of the scientific workforce.

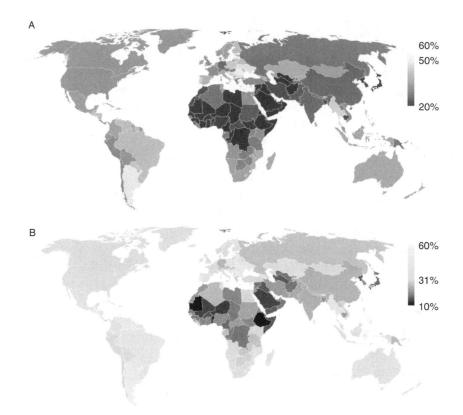

FIGURE 1.4. Proportion of women authorships by country, 2008–2020. Panel A presents the data with a color coding that emphasizes gender parity (50%). The darker the shade, the higher the dominance by men; the lighter the shade, the closer the country is to gender parity. Panel B presents the data with a color coding that emphasizes countries' relationships with the world average (30.9% women authorships). Darker shades denote countries that are below the world average in terms of their percentage of women authorships; lighter shades denote countries that are above the world average in terms of women authorships. Data from Web of Science core collection.

Many Middle Eastern and African countries—with Angola, Namibia, Mozambique, South Africa, and Tunisia as exceptions—can be found at the other end of the spectrum. Among those, Iran (23%), Jordan (22%), Bangladesh (20%), the United Arab Emirates (20%), Cameroon and Qatar (18%), Saudi Arabia (15%), and Ethiopia (10%) are worth noticing given their low rates of women authorship.[34] Farther east, several

Asian countries show lower than average proportions of women authorships: India (26%) and the Republic of Korea (21%) join Japan (17%) with low representation of women on the bylines of scientific articles.

As discussed earlier, percentages of authorships are a function of the number of women in the research system. If we examine the distinct number of women who have contributed to scholarly papers over the last thirteen years, instead of the proportion of women authorships, we see a slightly different picture. For almost every country, there is a higher proportion of distinct women authors—that is, of women who have authored at least one paper—than of their share of authorships (Figure 1.5). More specifically, while women account for 30.9% of authorships at the world level, they constitute 36.6% of distinct authors.[35] Therefore, most countries are closer to parity in terms of unique authors, and thirty have reached gender parity. Among the countries with at least 10,000 papers published between 2008 and 2020, those countries to have reached parity in women authors include Argentina (55%), Croatia (55%), Romania (54%), Portugal (53%), Thailand (53%), Tunisia (53%), Poland (52%), Slovakia (52%), Bulgaria (52%), Uruguay (52%), Lithuania (52%), Finland (51%), Brazil (50%), Czechia (50%), Estonia (50%), Serbia (50%), and Spain (50%).[36]

Geographic and political divisions in terms of women authors are particularly visible in Figure 1.5b, where we examine the percentage of women researchers compared with the world average (36.6%). European countries, as well as those located in North and South America, are all above average, while most countries in Africa and the Middle East, as well as China (29%), India (31%), and Japan (26%), are sizably below average. The gap between authors and authorship could be explained in two, slightly contradictory, manners: it could indicate a retention issue, where women enter the system but leave with few publications; or it could suggest that the research systems in these countries allow for the retention of women in science at lower levels of productivity.

Although the percentage of women as distinct authors is higher than the proportion of women authorships, these variables are strongly related: countries with a higher number of women in their research system as authors also have a higher percentage of women authorships (Figure 1.6, inset). However, there are exceptions to this, which shows that productivity levels by gender exhibit country-level variations. Among those countries with a sizable number of articles, Iran, Saudi Arabia, Germany, and Japan are notable examples, where the propor-

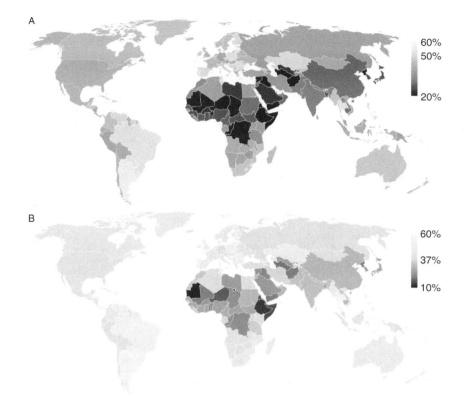

FIGURE 1.5. Proportion of women authors, by country, 2008–2020. This percentage is obtained by dividing the number of distinct women researchers who have authored a paper while affiliated to a country by all distinct researchers who have authored a paper affiliated to that country. Similar percentages are obtained when only researchers with more than one paper are included. Panel A presents the data with a color coding that emphasizes gender parity. The darker the shade, the higher the dominance by men; the lighter the shade, the closer the country is to gender parity. Panel B presents the data with a color coding that emphasizes countries' relationships with the world average (36.6% women authors). Darker shades denote countries that are below the world average in terms of their percentage of women authors; lighter shades denote countries that are above the world average in terms of women authors. Data from Web of Science core collection.

tion of women authors is 50% higher than their share of authorships. In other words, their presence in the workforce exceeds what would be expected given output alone. At the other end of the spectrum, the percentages of women authorships and distinct researchers are almost identical for China, South Korea, and Taiwan, which suggests that men

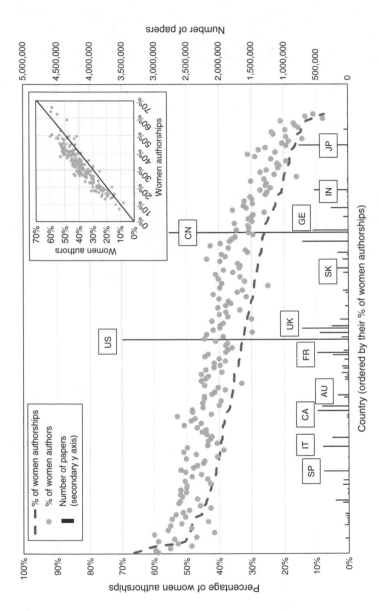
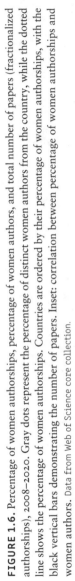

FIGURE 1.6. Percentage of women authorships, percentage of women authors, and total number of papers (fractionalized authorships), 2008–2020. Gray dots represent the percentage of distinct women authors from the country, while the dotted line shows the percentage of women authorships. Countries are ordered by their percentage of women authorships, with the black vertical bars demonstrating the number of papers. Inset: correlation between percentage of women authorships and women authors. Data from Web of Science core collection.

and women produce at rates equal to their representation in the work-force. These different cases suggest different policy interventions. In cases like Iran, Saudi Arabia, Germany, and Japan, the higher number of women may suggest strong educational systems but working climates that are unsupportive of women's employment in scientific fields. This reinforces the disconnection between educational and occupational access that has persisted in various countries across the past century. The equal rates of authors and authorship in China, South Korea, and Taiwan might suggest that an increase in women in the workforce will yield a concomitant rise in production.

The Interplay between Country and Discipline

Both country and discipline influence the participation of women in sci-ence. The interaction among these variables, however, is less understood. Does country specialization in a certain discipline yield higher degrees of women participation? Or does the presence of women in the labor force lead to country specialization in particular disciplines? To investigate this, we examine the participation rates of women at the intersection of disci-pline and country for the most productive countries (Table 1.1). We ob-serve a strong influence of country: those that have higher percentages of women authorships overall also have higher-than-average percentages of women authorships in each discipline.

Health is the only domain with higher proportions of women au-thorship overall. However, several countries overcome disciplinary gender barriers, with women's participation rates in other disciplines higher than the global average.[37] Portugal, for example, has higher rates of women's participation in mathematics than would be expected given the global average. In fact, several Spanish- and Portuguese-speaking countries (Portugal, Brazil, Spain, and Mexico) outperform what would be expected across many disciplines. The same is true for Asian coun-tries, with Taiwan being the most striking example. Despite these country distinctions, disciplinary cultures remain strong—see, for example, the low rates in engineering and physics across most countries. Overall, the data suggest a strong interplay between the scientific and cultural di-mensions in predicting rates of participation by women. That is, gen-dered production in science is highly contextualized and situated in the intersectional space of discipline and country. Policies to alleviate gender

TABLE 1.1. Percentage of women authorships, by country and discipline, 2008–2020. Thirty countries with the most papers over the period. The color coding emphasizes countries' relationship with the world average, all disciplines combined (30.9% women authorships). The darker the shade, the higher the dominance by men; the lighter the shade, the closer the country is to gender parity. Data from Web of Science core collection.

Country	Health	Psychology	Arts	Professional fields	Humanities	Clinical medicine	Biomedical research	Social sciences	Biology	Chemistry	Earth and space	Engineering	Mathematics	Physics	All disciplines
Portugal	57	56	46	44	44	50	55	39	53	47	45	25	34	22	44
Poland	58	55	48	42	45	51	55	44	53	48	47	27	21	24	44
Czechia	54	51	46	40	31	54	52	39	47	38	41	32	23	36	42
Finland	68	56	60	52	48	48	46	48	44	35	39	24	16	20	42
Brazil	68	52	41	34	35	47	49	33	38	39	36	23	16	16	41
Spain	53	49	40	45	45	45	46	38	44	39	38	26	23	21	40
Taiwan	52	47	49	44	49	41	43	41	43	37	38	36	36	34	39
Italy	49	52	44	37	36	41	49	33	42	41	33	23	29	17	38
Australia	66	52	50	47	44	41	37	40	30	24	26	18	17	16	37
Sweden	66	42	52	43	39	42	38	38	35	27	32	18	17	16	36
Israel	62	53	53	53	37	39	41	36	34	31	25	22	17	17	35
Mexico	52	48	36	36	41	42	43	34	34	35	33	21	19	17	34
Netherlands	52	46	42	36	36	37	35	31	27	21	25	17	17	17	33

Canada	63	51	50	42	41	39	35	38	33	23	26	16	16	15	33
United States	61	50	48	43	40	36	34	35	31	23	25	18	18	16	33
Denmark	58	47	43	37	36	38	35	32	32	22	26	15	13	12	33
France	47	44	45	34	40	38	42	33	36	31	29	21	18	18	32
United Kingdom	56	46	44	39	38	36	35	35	30	22	24	16	16	14	32
Turkey	62	51	52	38	37	34	39	34	32	36	30	21	27	19	32
Belgium	47	42	42	38	36	36	37	32	30	26	26	19	18	16	32
Singapore	54	48	33	36	31	38	34	30	33	27	30	22	19	22	29
Russia	48	58	55	46	38	51	46	38	43	35	29	21	15	15	28
Austria	47	43	44	38	36	32	35	33	32	23	23	14	15	12	28
Switzerland	46	41	36	30	35	31	32	31	29	23	24	17	16	14	27
Germany	43	43	35	32	34	30	34	30	33	22	24	15	15	14	27
China	38	34	28	29	27	30	29	26	29	28	25	22	25	23	26
India	37	39	45	28	40	31	32	34	29	24	26	21	22	23	26
Iran	37	34	33	21	25	32	31	20	23	28	19	14	17	19	23
South Korea	43	35	28	25	30	25	27	22	26	21	20	13	16	13	21
Japan	34	27	28	23	30	19	22	18	22	16	16	11	11	10	17
All countries	57	47	43	40	37	35	35	34	33	28	27	20	20	18	31

disparities, therefore, must take both discipline and country as strategic organizational sites.

Economic and Political Factors

One example of the strength exerted within a country can be observed in Iran. Our results show that Iranian women are relatively underrepresented in every discipline. Yet this is not a consequence of a lack of capacity. Take, for example, Maryam Mirzakhani, the only woman and Iranian to win the Fields Medal for excellence in mathematics. That she was extraordinary is undoubtable. There may, however, be several young women who do not receive a chance to excel. Girls score much higher than boys on science and mathematics in primary school in Iran and across other countries in the region. Saudi Arabia has the highest gap in favor of female achievement in grade 4, with Bahrain, Oman, Kuwait, Qatar, and the United Arab Emirates following.[38] In Iran, the female achievement gap continues in grade 8 but is smaller. By grade 12, the differential is lost, although women who take college entrance exams enter at higher rates than men (42% of women compared with 29% of men).[39] This leads to a slight overrepresentation of women among undergraduate students (51%), but attrition is quickly observed. Women represent only 28% of researchers, slightly below the global average.

Political pressures explain some of these differences. Mirzakhani studied mathematics in Iran at the undergraduate level before going to Harvard University for graduate school. Systematic barriers, however, prevent other women in Iran from achieving a similar trajectory. For instance, in 2012, thirty-six Iranian universities banned women from seventy-seven different fields of study.[40] This made disciplines such as accounting, engineering, and chemistry only available to Iranian men. At the University of Tehran, Mirzakhani's field of mathematics moved to exclude women. This decision was purely ideological, rather than empirically based, reinforcing the role that geopolitical factors play in creating global gender disparities.

At the macro level, other geopolitical factors can be observed through economic and development data, such as those available from the World Bank.[41] Figure 1.7 presents several relationships between country-level variables and the proportion of women as authors of scholarly papers. There is, unsurprisingly, a linear relationship between women in the labor force, generally, and in science, specifically. The relationship is particularly strong for high-income countries and speaks to the ability of women to leave the domestic sphere and enter various sectors of society. However, there is a slightly negative relationship between women's participation in science and gender differences in advanced education. In almost all countries, a higher proportion of women obtain advanced education (university degrees), yet the proportion of active researchers remain lower than that of men. This suggests that advanced education alone will not diminish gender disparities in science. Similarly, political empowerment—which is measured through the percentage of women in seats in national parliaments and the percentage of ministerial positions—has little relationship to women's proportion of scholarly papers (Figure 1.7).

Furthermore, as highlighted in the introduction, higher percentages of women in the workforce may be a sign of economic destabilization, as is the case for many Soviet countries.[42] For example, participation of women in science is positively associated with men's unemployment. In a strong economy, we might expect women to enter into the labor force at similar rates to men; however, in less robust economies they are more likely to fill a vacuum left by men who are unemployed or where scientific work is not associated with high degrees of social and economic capital. The most striking evidence of this paradox is the relationship between life expectancy and women authorships, the strongest correlation in our analysis. In countries where men die younger, there are more women in the scientific labor force, suggesting that substitution and brain drain effects are present.[43] The countries with the highest proportion of women authorships are marked examples of this. In Latvia, men's average life expectancy sits at 68.8 years, whereas women live, on average, an additional 10 years (78.8).[44] The same gap is observed in Ukraine: women's life expectancy at birth stands at 75.8 years, whereas men's life expectancy is 65.5 years. When there is an absence of men (through either war or mobility), women are disproportionately represented in science.[45]

Another factor that seems to have a positive (though relatively weak) association with women's participation in science is the number of paid maternity days. The provision and reporting of leave days varies across countries; however, among those that provided information, we can observe that countries with longer paid maternity leave (mostly former Yugoslavian countries) also have a higher proportion of women authors.

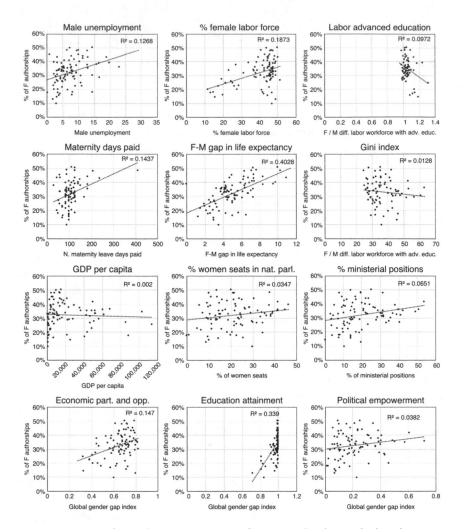

FIGURE 1.7. Correlations between percentage of women authorships and selected country-level indicators of economic and social development (World Bank Open Data). For countries with more than 500 fractionalized authorships over the 2008–2020 period.

However, once these countries are removed, the relationship dissipates. This can be linked with the paradoxical nature of parental leave: the more generous the leave is, the longer the researcher will be out of the field and, therefore, the less likely he or she is able to contribute to scholarly articles.

Several other economic and gender-related variables seem to have an insignificant relationship to women's participation in scientific activities and, in some cases, a negative association. For instance, indicators of economic development (GDP per capita) and concentration of wealth (Gini index) have little relation with women's scientific production: countries that are more developed and where income is more equally distributed do not exhibit greater gender parity in science.[46] Inversely, there are many countries in southeast Europe, the Caribbean, Latin America, and central Asia with lower levels of scientific infrastructure and high concentrations of wealth that are achieving parity.

Our previous work suggests that these relationships—between parity in production and indicators of human development and gender equity at the country level—are not linear.[47] In fact, those ranking the highest on human development indicators often have less gender parity in production than those in the lower category, with the greatest equity observed in the countries ranked in the middle. It is necessary to repeat the warning of sociologists J. Scott Long and Mary Frank Fox here: "It is not enough to ask whether particularism of universalism operates in science . . . Rather, given the patterns observed, it is important to understand more fully the processes leading to the lower participation and performance of women and minorities in science."[48]

A Parity Paradox

Several studies suggest that trends toward parity in production are, at least partly, driven by brain drain: among the countries with a higher share of women authors are also those where a large proportion of scholars have left the country in order to access better resources in other countries.[49] Our analysis of mobility suggests that men are likely to have greater degrees of mobility than women; therefore, in cases of brain drain, women are likely to be the ones left behind.[50] This is one of the cases where we observe a "parity paradox"—a growing rate of gender parity in a country does not necessarily indicate that this country has achieved greater equity. In fact, it may demonstrate the reverse; gender parity in scientific production may have equalized due to greater opportunities for men in other sectors within the country and to scientific opportunities outside the country.

This is exemplified by Poland, one of the only countries with substantial production and parity in the scientific workforce—women represent

51.8% of publishing authors, although they still lag in terms of author-ships (42.5%). Poland has a strong history of women's participation in the workforce: by 1970, women formed half of the workforce for all eco-nomic sectors combined; by 1980, 60% of medical students were women. Women occupied a sizable proportion of jobs in finance, health, and edu-cation and even ran 20% of farms.[51] However, the rate of women in these sectors came at the expense of women in politics and government, where there was a "male monopoly on power."[52] Communism espoused princi-ples of equality; therefore, turning to a market economy led to greater discrimination among women, as it was associated with a "revival" of the traditional role of women.[53] Despite Poland's election of its first woman prime minister in 1992, there has been a regression of women's rights in recent years. In 2017, the police raided women's rights associations, and sexual rights (such as sex education and abortion) have diminished.[54] Po-land's regression is an example of the parity paradox, where equity has eroded although parity remains.

Poland's story is not unlike that of its northern neighbor, Latvia, and several other former Soviet states. Latvia has a high proportion of women researchers (60%) and a long history of gender equity—in terms of ac-cess to education, employment, and equal pay.[55] Following the fall of the USSR, several conditions accentuated the entrance of more women into the workforce. The first were the harsh economic realities of former So-viet countries, where a single breadwinner could no longer sustain a household, which led to an increased proportion of women seeking work.[56] However, there was another, more morbid reality, as discussed earlier. Across all countries—and strikingly for Latvia—one of the stron-gest predictors of a higher participation of women in science is a gendered difference in life expectancy.[57] The fact that most university students are women and women have low unemployment may not necessarily be a positive indicator for the country, but rather a reaction to the lower vol-ume of active men within the society.[58] This overbalance toward women has led to a devaluation of labor in Latvia: women make roughly 25% less than men in Latvia, and academic salaries are among the lowest in Europe.[59] This follows several global examples of depressed salaries for woman-dominated occupations.[60]

Gender parity in Poland masks a clear economic and social segrega-tion: women have had equal access to education and relatively equal par-ticipation in research for decades. However, they suffer from higher un-employment rates, are paid approximately 30% less, and perform a higher proportion of domestic duties.[61] Therefore, this parity does not constitute equity. Rather, parity in the scientific workforce reflects feminization of a sector to which a lower symbolic value is given. This relates to the social and economic capital associated with men's work: politics and government were held in higher regard; therefore, men concentrated in these areas, leaving academe open for women. The same can be observed in disciplinary specialization, with women appearing in disciplines that are at the lower levels of the hierarchy of disciplines and receiving lower pay, even in highly feminized disciplines.[62] Parity in science is a valuable goal but one that should not be pursued at the expense of equity.

Of course, equity is unlikely to occur without parity. Therein lies the paradox. Those of a more quantitative bent will focus on the numbers and rejoice when these numbers hit their quota. There is reason to celebrate achieving parity; however, we argue that although necessary, it is not a sufficient criterion for mitigating disparities in science. Parity can often mask underlying inequities. The goal of this book, therefore, is to bring to light as many of the dimensions underlying these differences as possible. As data scientists Catherine D'Ignazio and Lauren F. Klein have argued, "Counting and measuring do not always have to be tools of oppression. We can also use them to hold power accountable, to reclaim overlooked histories, and to build collectivity and solidarity."[63]

Measurements of production can and have been used to exacerbate inequities in the scientific workforce. Therefore, we use them with caution, to document disparities, but with a careful critique of parity in the absence of equity. The following chapters seek to dig deeper into these numbers, to occasionally leave the numbers and allow the women to tell their own stories, and to weave in the historical, sociological, and economic factors that are essential for understanding how we can move toward equitable parity in the scientific ecosystem.

Chapter 2

Collaboration

Collaboration is a hallmark of contemporary science and served as an entryway for many women. However, it can come at a cost. Lillian Gilbreth (1878–1972), for example, is often presented as the quintessential example of a woman who balanced a collaborative career and family life. Born Lillie Evelyn Moller to a wealthy family in California, she had an aptitude for school and desire to attend college that caused her family consternation, as they felt this trajectory was not appropriate for a woman of her status. Undeterred, she convinced her parents to allow her to attend Berkeley and study English. After graduation, she changed her name to Lillian and moved to New York to study philosophy and comparative literature at Columbia, where lectures were open to women with the approval of a faculty member. Lillian attended a few courses in psychology before returning to Berkeley for a graduate degree in Elizabethan literature. She then began a doctoral program in psychology at Berkeley but took a semester off to travel in Europe. It was on this trip that she met Frank Gilbreth—the cousin of her chaperone. Frank had taken the MIT entrance examinations but decided not to go because it would financially burden his mother, who was fully supporting the family. He began working in construction and, by the time he met Lillian, had one of the largest construction companies in the United States.[1] They married in 1904 and relocated to New York City.

Despite completing her dissertation in 1911 at Berkeley, Lillian was not awarded the degree because of her failure to meet the program's residency requirements. Frank convinced Brown University to admit Lillian into their doctoral program on the conditions that she retake her oral examinations and write a completely new dissertation. She finished in two years, with a thesis that focused on increasing efficiency in school classrooms. Therefore, by 1915, Lillian had written two dissertations and had seven (of eventually a dozen) children. As Brown University president W. H. Faunce wrote to Frank upon graduation, "I do not know another woman in America who has achieved what she has done in the realm of study, and at the same time fulfilled every duty of motherhood in her constantly enlarging home."[2]

Lillian wrote extensively with her husband, though her name was often absent. They coauthored the famous *Primer of Scientific Management* (1912), but the publisher refused to put Lillian's name on the cover.[3] The same was true for *Concrete Systems* (1908), *Bricklaying System* (1909), and *Motion Study* (1911), in which the Gilbreths described their "One Best Way" system. She also edited an important volume in management and wrote many of the papers for Frank's presentations. Frank tried to have Lillian's thesis published, but it received no attention. Finally, as other scholars began to publish similar ideas, he persuaded the journal *Industrial Engineering* to publish some sections of the thesis, under the name L. M. Gilbreth. Macmillan soon agreed to do the same, publishing the entire thesis as L. M. with no mention of the author's gender in the initial printing or the reprints in 1917 and 1918.

When Lillian was inducted as the first woman honorary member of the Society of Industrial Engineers, Frank joked that his own success was the result of the "sweat of his frau." Despite this, Lillian often appeared as "helpmeet to a more qualified scientific husband" rather than an equal intellectual partner. When Frank unexpectedly died in 1924, they were about to leave the country to attend the World Power Conference in England and the International Management Congress in Prague, where he and Lillian would be representing the American Management Association. Lillian went alone, presided over the sessions, and presented on the Gilbreth brand of "motion study." However, the world of industrial engineering was not ready to have a woman at the helm; contracts were canceled or not renewed and meetings were held at venues where she was not allowed, such as the University Club or the Engineers Club.[4] She quickly pivoted to fields where her status as a woman was less threatening: she began to teach industrial psychology to firms and consult for retailers, and she turned her research to domestic spaces, such as the kitchen and family. The "One Best Way" was now applied not for concrete or bricklaying but in *The Home-Maker and Her Job* (1927) and *Living with*

Our Children (1928). Whereas previous women had to choose between science and family, Gilbreth found an avenue for research by marrying the two.

Another academic couple in psychology was Mamie Phipps Clark (1917–1983) and her husband, Kenneth Bancroft Clark. Both were African American scholars with doctorates from Columbia. As with Lillian, Mamie's work focused on the domestic sphere. In particular, she was concerned with the provision of psychological services for Black children. Clark was born in 1917 in Hot Spring, Arkansas, and her graduation from high school was met with several scholarship offers for college, including historically Black institutions such as Fisk and Howard Universities. She chose the latter and started as a math major (with a minor in physics). Her passion was dampened after a year at Howard, where there was little support for women math students.[5] She switched to the psychology program.

It was at Howard that she met her future husband, Kenneth, who was a master's student in psychology. She joined him in the graduate program in 1938. Mamie's graduate thesis marked the beginning of her line of research on race consciousness, which had a historic impact on the desegregation of American public schools. Mamie started a doctoral degree at Columbia and began collaborating with her husband on what became known as the "Dolls Test."[6] Kenneth acknowledged that this work—which was cited by the Supreme Court in the 1954 *Brown v. Board of Education* decision—was primarily Mamie's and that he "sort of piggybacked on it."[7] Unlike in the Gilbreths' publications, where Lillian's name was omitted, in the Clarks' work, Mamie's name stood on the byline with Kenneth's. Despite this, it was Kenneth, not Mamie, who was asked to give expert testimony and wrote the analysis and expert report for the landmark case.[8]

The Clarks' academic outcomes continued to be unequal. When she graduated in 1943, Mamie was the first Black woman to earn a doctoral degree in psychology—and the second Black person after her husband. Kenneth found a teaching position at the City College of New York, but Mamie was unable to find a position. Therefore, she renovated a basement in Harlem (with funds from her family) and opened the Northside Testing and Consultation Center, later renamed the Northside Center for Child Development.[9] This facility provided services that were unavailable at the time for children from minoritized populations. Many Black children were being incorrectly diagnosed with disabilities within the public school system and removed from the classroom. To address this, Mamie provided free remedial classes; this was about not only child development but also civil rights.[10]

Mamie served as the director of the center from the founding (in 1948) until her retirement in 1979. Her legacy lives on in each of the children and the families that she served during these decades (and the continued work of the center). However, the inequalities in recognition compared with those of her husband are notable. While they shared awards from Colombia University (such as the Nicholas Murray Butler Silver Medal and the Distinguished Lecture Award), there were several prestigious awards that went only to Kenneth for his contribution to *Brown v. Board of Education,* including the Spingarn Medal from the NAACP (1961), the Presidential Medal of Liberty (1986), and the American Psychology Association Award for Outstanding Lifetime Contribution to Psychology (1994). Their collaboration yielded important contributions to science and society, but he reaped most of the accolades.

Darshan Ranganathan (née Markan) (1941–2001) also constituted one side of an academic couple, but she was able to slowly uncouple her academic career from her husband's and receive credit for her own scientific work. Darshan was born in Delhi in 1941. She received a PhD in chemistry from Delhi University in 1967 and was hired as lecturer and then head of department at Miranda College. Quickly thereafter, she received a research fellowship to do postdoctoral work at Imperial College London. She returned to India in 1969 and, in 1970, attended the Indo-Soviet Binational Conference on Natural Products, where she observed a talk by Subramania Ranganathan, and a relationship ensued. They were married that year, and she joined Subramania at the Indian Institute of Technology, Kanpur (IIT Kanpur), as his research associate.

Darshan was supported at IIT Kanpur by a series of fellowships, as there were (and remain) unwritten rules that spouses could not be hired as faculty in the same department.[11] The objective of the practice is to avoid conflicts of interest and nepotism. In this case, however, it stymied the development of a scientist. As one colleague wrote, "I will always harbor the sore point in me that Darshan was not given the credit and the position that she truly deserved early enough. Despite her history of achievements and ongoing activities, she was never considered for a faculty position, while lesser colleagues rose to become Professors and Vice-Chancellors."[12] Darshan's husband sought to correct this, stating, "I knew from the beginning that she was better than me and was proud to share my funds and students with her so that she could work on her own problems and publish on her own. That was all she wanted, brushed away all other irritations and slowly blossomed into an organic chemist who won international peer recognition."[13]

The consequence of her position led to a deep level of collaboration between the couple. As one biographer wrote, the move to IIT Kanpur "was the start of a legendary period in their twin careers, each supporting and enhancing the other to reach greater levels of creativity and achievement." Often cited is their collaboration as editors on *Current Highlights in Organic Chemistry,* as well as their coauthored books and journal articles. In contrast to the academic couples discussed earlier, the contribution of the wife was not dismissed in later obituaries. As one man concluded, "Their lives became so intertwined that to think of one without the other became impossible."[14]

Upon her husband's retirement in 1993, Darshan moved to the Regional Research Laboratory in Thiruvananthapuram (formerly Trivandrum). She was there for five years until she took a position as deputy director at the Indian Institute of Chemical Technology in Hyderabad. One biographer noted that, despite her previous productivity with her husband, these independent moves were transformative in her career. An analysis of her work demonstrates this quantitatively: in 1970, all her publications were with her husband; this dropped to 56% in the 1980s and to 12% in the 1990s. None of her publications in the 2000s were written with her husband. Her work took different paths as she reached out to new collaborators. One important collaborator was Isabella L. Karle of the Laboratory for the Structure of Matter at the Naval Research Laboratory in the United States. The two women collaborated for more than seven years, publishing two dozen papers together without ever meeting. It was during this time that Darshan, as one biographer wrote, "shone in collaborative glory."[15]

Unlike Mamie and Lillian, Darshan received recognition for her work. She was elected to the Indian Academy of Sciences in 1991 and the Indian National Science Academy in 1996. She was awarded the Third World Academy of Sciences Award in Chemistry in 1999 and received the Jawaharlal Nehru Birth Centenary Visiting Fellowship in 2000. This last fellowship came three years after she was diagnosed with breast cancer, and it was her last trip abroad. Her productivity continued until the end, with several works published posthumously. As one obituary asked, "Who amongst us can match her record of 11 papers in the *Journal of the Amer-*

ican Chemical Society and 6 papers in the *Journal of Organic Chemistry* during the last five years?"[16] As her husband poetically summarized, "She was a comet on the chemical horizon, shedding brilliance at prodigious costs of energy and vanishing at the apex of her career."[17]

Like the women before her, Darshan battled the politics of distinguishing herself within a collaborative partnership, but she eventually emerged as a recognized scientist in her own right. In these three cases, spousal collaborations demonstrate how time, place, and race have led to differential outcomes. However, the growing ubiquity of collaboration in science requires a more comprehensive analysis of the costs and benefits for women in this scientific practice.

From Authorship to Collaboration

There have been, throughout history, several instances in which religious, political, racial, gender, and other social politics prevented scholars from claiming a publication as their own.[18] Researchers—both men and women—took advantage of anonymity and pseudonymity at various times to engage in criticism or to avoid association with controversial or heretical material. For women, it was sometimes the only way to enter scientific publishing.[19] Initials have also been used to mask gender on a publication, at the insistence of the author or, in Lillian Gilbreth's case, the insistence of the publisher.[20] Anonymity, however, has been replaced by a hyperfocus on credit, as a response to the rise of research evaluation systems at the end of the twentieth century. Authorship allows for the attribution of credit as well as the assignation of responsibility for the underlying work; it serves as an indicator of one's contribution to scientific work.[21] The scientific reward system pivots on this notion: authorship denotes contribution to a piece of work, which is rewarded with the visibility associated with the publication, as well as productivity and impact metrics that are increasingly assigned to scholarly publications.

The conceptualization of authorship assumes both a shared understanding of the capital that authorship provides and an implication that authorship indicates contribution to the work that is presented. Historical notions of authorship—that is, penning the words of the articles—align with this conceptual model. However, the model has been challenged in recent years with changes in both research practices and authorship conventions.[22] Scientific collaboration has increased steadily over the course of the twentieth century. In 1903, when academic couple Marie and Pierre Curie were jointly awarded the Nobel Prize for Physics, 85% of publications in the natural and medical sciences were written by a single author. Today, that number is less than 10%.[23] The hyperbolic nature of authorship hit the news cycle in 2015 with a publication in high-energy physics by the CERN team in Switzerland that had 5,154 authors, setting the record for the largest number of contributors to a single article.[24] The publication was thirty-three pages long, with twenty-four pages dedicated to listing authors and their institutions. Only nine pages were devoted to the actual research. CERN broke its own record in 2019, publishing a piece with 5,216 authors.[25] While much less common, such large collaborations are also present in medical research. For example, a paper on the prevalence and treatment of cardiovascular diseases published in the *Journal of the American Medical Association* listed, in addition to the 11 main authors (all men), every investigator in the REACH Registry, which included 5,576 contributors.[26] In that case, the paper was eighteen pages long, with the last eight pages devoted to the list of additional contributors.

One could dismiss these trends as being strictly due to the fact that in some subfields, research revolves around large infrastructures, such as particle accelerators or observatories.[27] This, however, is not the case. We see increases in collaboration occurring across all fields. Even in arts and humanities, where the author is still seen as the (often sole) person who writes the manuscript, transformations are happening. For instance, while less than 10% of publications in the field were collaboratively authored in 2005, that rate is now around 20%. More striking are the social sciences, where rates of collaboration went from less than 40% in 1980 to nearly 73% in 2013.[28]

Understanding collaboration patterns by gender is complementary to production analysis, as it provides insights on the context in which research activities are conducted. The increased number of authors per paper observed in all disciplines is due, in part, to the growing complexity and interdisciplinarity of research, but it also demonstrates a change in credit attribution practices and increasing recognition of various roles, such as those performed by "invisible technicians" in science.[29] By placing the names of all those who contributed to a piece of research on the byline, we acknowledge that authorship, the coin of the academic realm, is a currency that should be provided to all those who labor in science—student, technician, and established researcher alike. As we will show in this and the following chapters, women compose a substantial portion of these invisible scientific workers.

Despite the centrality of authorship, there is wide disagreement over the concept and implementation of authorship—that is, who is named on a paper and the order in which they are placed. First, there are considerable differences by discipline, in both author naming and ordering.[30] For instance, in disciplines of the social sciences and humanities, the act of writing remains central to authorship—only those who have contributed to this task will have their names on the byline. On the other hand, the disciplines of the natural sciences and medical sciences recognize a much wider spectrum of contributions, with writing being one of many. Therefore, the number of authors on a publication—the main indicator of team size—varies drastically by discipline, with the humanities having the smallest teams and natural sciences the largest.[31]

Alphabetical listing was common until the mid-twentieth century and has been employed more recently in fields such as mathematics and economics.[32] This practice, however, is vanishing: for articles indexed in Web of Science in 2011, less than 4% included alphabetical listing, although this percentage was above 50% in four disciplines (mathematics, finance, economics, and particle physics).[33] As authorship lists become longer, there is greater emphasis placed on first and last authors, as these are typically the dominant authorship roles, associated with the highest levels of contribution.[34] High-energy physics works on a slightly different model, using the "standard author list" as the model. For example, the Collider Detector at Fermilab Collaboration created several rules for becoming an author: one must dedicate a portion of time (50% for academics and 100% for graduate students) for three years in order to be placed on the standard authors list; after this time, one's name is placed on every publication emerging from the collaboration.[35] The list is updated biannually to include all those who qualify, meaning that the overlap between labor and publication may not be precise—that is, one may contribute to a publication but not be on the author list or vice versa. This system pays in academic capital for physical labor and removes "responsibility" from the equation by accounting for the contribution of labor without a direct link to an intellectual product. As historian of science Mario Biagioli anticipated two decades ago, "Like credit, responsibility appears to be turning into a more operational category and less of an essential feature attached to the name of the author. This turn toward operational views of credit and responsibility seems to be coupled with an increasing subdivision and distribution among different people of the functions that used to be kept together under the all-encompassing figure of the author. Scientific authorship as we knew it may be falling apart, or it may be simply unburdening itself of all those functions it could no longer juggle together."[36]

The lack of an explicit relationship between labor and authorship has many consequences for the knowledge economy. First, there have been gross abuses in terms of fraudulent authorship practices.[37] Along with honorary authorship, omission is one of the largest forms of authorship malpractice.[38] Ghost authorship is the failure to list an author on the byline who has made substantial contributions. For the biomedical fields, the proportion of papers with ghost authorship hovers around 10%.[39] In the strictest sense, ghost authorship is seen as an intentional and malicious malpractice. However, omitting those who played significant roles is commonplace: in a recent survey, more than half answered in the affirmative to having a researcher or student who played a significant role in the paper but was omitted from the coauthor list. This may reflect differing norms of authorship across disciplines (and within disciplines for authors of varying status). In a review of ghost authorship in industry-initiated trials, for example, nearly all ghost authors were statisticians.[40] This might suggest that ghost authorship is not always malicious but reflects varying criteria for authorship across domains. Simply put, statistical work may not rise to the level of authorship in industry publications. This is not a novel phenomenon—technical and analytic work has been historically discredited in the hierarchy of scientific labor.[41] As we will discuss in more detail later, this perpetuates inequalities in the scientific workforce, as this labor role is disproportionately associated with graduate students, technicians, and women—and these individuals are least likely to garner credit for this work. As management scholars Carolin Haeussler and Henry Sauermann note, "Contributions in the forms of carrying out technical steps or laboratory work are more likely to be rewarded with authorship when made by scientists with higher hierarchical status."[42] That is, when tasked with the same technical work, senior authors are more likely to be placed on the byline than junior scholars (who are disproportionately women).

Honorary authorship falls on the other side of the spectrum—providing authorship for those who did not make a substantial contribution—and has been found to be more likely associated with review articles.[43] This has several explanations. It could suggest that review articles—where the labor is predominantly that of "writing"—provide more ambiguity for author-ship than more empirical articles. It could also suggest that these publica-tions are not valued in an equivalent way—allowing for honorifics deflates the value of authorship on these articles. However, a third argument could be made that review articles need a certain cachet for publishing and that these honorific authorships confer the necessary prestige. At any rate, it suggests looser criteria for this labor role than for others. Honorific au-thorship is related to the malpractice of "salami slicing," where authors seek to extend their research over as many publications as possible. In both salami slicing and honorific authorship, the incentive for the author is to increase units of production. In a Gilbreth-type approach to labor man-agement, the most efficient model is one in which they gain authorships without a large investment in labor. This explicit link between labor and credit—without reference to a particular product—is made manifest in the Collider Detector at Fermilab Collaboration agreement.

The practice of equating labor with publications places a strong em-phasis on authorship: appearing on a byline, and the order in which one does, serves as a critical signal of labor. However, these explanations fail to distinguish between disciplinary differences in value ascribed to various scholarly tasks. One way that has been used to examine this is to look at the relationship between authorship, those who appear on the byline of a paper, and subauthorship, those who are found in the acknowl-edgments field.[44] In an analysis of funded research in Web of Science, it was found that differences in the number of authors are reduced when subauthors (those listed in the acknowledgments) are included.[45] For example, the average number of coauthors in the social sciences is slightly below three, whereas it is above five for chemistry. However, when sub-authors are included, there is a nominal difference, with both averaging between five and six contributors. This may suggest that it takes the same number of authors to produce a work across these two domains, but while nearly all are given authorship in chemistry, only about half receive this level of credit in the social sciences.[46] Interestingly, how-ever, the gender disparity in authorship attribution is also found in acknowledgments, with women being even more underrepresented in acknowledgments than in authorships.[47] All of these findings clearly suggest that there are distinct disciplinary differences in how authorship is used to provide credit for scientific labor, and that those are compounded by gender.

In 1988, the International Committee of Medical Journal Editors attempted to address these concerns with a statement on specific requirements for the attribution of authorship. This statement provided three criteria that must be met for someone to be considered an author on an article: "1) substantial contributions to the conception or design of the work; OR the acquisition, analysis, OR interpretation of data for the work; AND, 2) drafting the work or revising it critically for important intellectual content; AND, 3) final approval of the version to be published; AND, 4) agreement to be accountable for all aspects of the work in ensuring that questions related to the accuracy or integrity of any part of the work are appropriately investigated and resolved."[48]

The and/or formulation of this statement is important: authors should have done both the writing and one of the following: conception or design, data acquisition, or analysis. Furthermore, they must all approve the final version and agree to be held accountable for all aspects of the work. Despite these concerns, fraudulent authorship practices have not been curbed. Furthermore, these requirements did not address the distance between labor and authorship in an era when scientific work is highly specialized and distributed. To understand the relationship between the organization of team science and the characteristics of those who labor, we overlay gender on the space of scientific production, with a focus on collaboration.

The Gendered Nature of Collaboration

The rarity of sole authorship in most specialties of the natural and medical sciences is particularly striking for women. In many disciplines, writing a paper alone is a marker of either a more theoretical contribution— which is often perceived to be of higher value—or a review-style contribution that synthesizes the research made on a given topic, which is known to attract higher citation rates.[49] It is in those disciplines where sole authorship is the most unusual that we see the strongest gender differences. Men are 60% more likely in biomedical research and biology to author papers alone; 50% more likely in physics, clinical medicine, and psychology; and 30% more likely in chemistry, engineering, mathematics, and earth and space sciences. This suggests that, overall, women's authorships are more likely to be in collaboration than men's and that women are less likely to be sole author in domains where single authorship is particularly distinctive or by invitation only.[50]

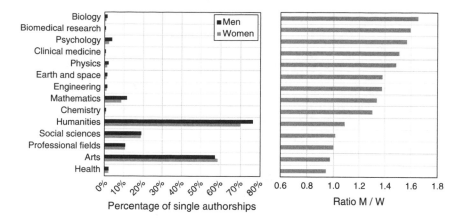

FIGURE 2.1. Percentage of articles that are single authored (left panel), and men-to-women ratio of percentage of single authorship (right panel), by NSF discipline, 2008–2020. Percentages of single authorships are obtained by dividing the number of single authors (or number of papers with single authors) by the sum of all authors on all scholarly papers. Data from Web of Science core collection.

Contemporary science, however, is a fundamentally collaborative endeavor. Collaboratively authored articles became the norm in scientific disciplines after World War II and in the social sciences by the end of the 1980s (Figure 2.1).[51] Only articles in the arts and humanities remain largely single authored. Even with this norm, there remain gender differences in who writes single-authored papers, with men having higher representation in the humanities and women in the arts. These disparities have implications for reward systems where single authorship is either privileged or required.

Disparities continue to exist across dominant author positions—that is, first and last author. There are no author positions where women are in the majority, but they are most likely to be found earlier rather than later in the author list. More specifically, all disciplines combined, women represent 35% of first and 33% of middle authorships, but only 25% of last authorships. Higher representation in first authorship is often associated with more-junior scholars, and women have greater representation among early-career researchers than they do among senior researchers.[52] In terms of middle authorship, however, their higher percentage is a reflection of the more technical tasks they perform, which have lower levels of prestige and recognition.[53] The most senior position—that of last

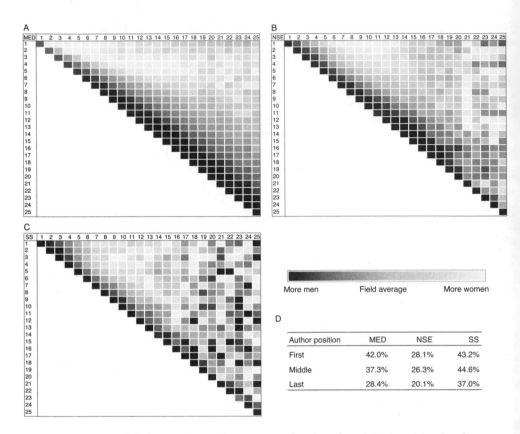

FIGURE 2.2. Relative percentage of women, as a function of team size (*x* axis) and author position (*y* axis), 2008–2020. Percentages are compiled within field. Darker cells indicate that men are more likely—compared with the field average—to be at that author position. Lighter cells indicate that women are more likely to be at that author position. Panel A: Medical sciences (MS). Panel B: Natural sciences and engineering (NSE). Panel C: Social sciences (SS). Panel D: Percentage of women authorships, by author position and domain, for papers in collaboration. Data from Web of Science core collection.

author—is strongly associated with men. This remains true irrespective of team size (Figure 2.2, panels A–C).

There is, however, an interesting pattern that emerges as the size of a team increases. Controlling for field differences, we find that women are more likely to be first authors as team size increases, until the team reaches about a dozen researchers (exceeding the mean values for most of the fields), after which this proportion decreases. Women also cluster more as initial authors (in the first third of authors listed on the byline) compared with men, who are featured at the end of the authorship list. These patterns evoke differences both in seniority and in types of contributions that warrant authorship, with strong implications for careers. As we will describe in Chapter 3, women are more likely to receive credit for writing and technical work (thereby placing them earlier on the byline), whereas men are more likely to receive authorship for the contribution of resources (placing them later on the byline). All of this speaks to the differential labor expectations for women and how the heterogeneity of middle authorship may obscure contributions in collaborative work.

Team composition, however, is highly influenced by dominant authors. To better understand how the gender of lead authors affects team size, we compiled the distribution of articles by number of authors and gender of those in leadership positions (Figure 2.3). This creates four categories, by first and last author combinations: M-M, M-W, W-M, and W-W. The distribution is examined in two different manners: within leadership type (horizontally, x axis; panel A) and within a certain number of authors per article (vertically, y axis; panel B). The leadership comparison shows minor differences in the relative percentage of team sizes: across all domains, publications with men as first authors and women as last tend to have fewer authors than the other compositions. For natural sciences, engineering, and social sciences, articles with mixed-gender leadership (M-W or W-M) are associated with smaller teams. However, the difference in team size by gender of leadership remains minimal, except in social sciences—which is likely a consequence of the lower proportion of articles with a high number of authors.

Unsurprisingly, articles with men as first and last authors are the most common mode of production in all fields, and the likelihood of having men in those two leadership roles increase with team size (Figure 2.3, panel B). This is particularly true in medicine, where larger teams are less likely to have women as last authors. Articles with women as first authors and men as last authors are slightly more prevalent for midsize collaborations and become less frequent as team size increases. Papers with women in last-author positions are the least common in all domains and are particularly uncommon in natural and medical sciences, with the lowest frequency being those with women in both first and last. Taken globally, across all domains, the results show that the larger the team, the less likely it is to be led by women.

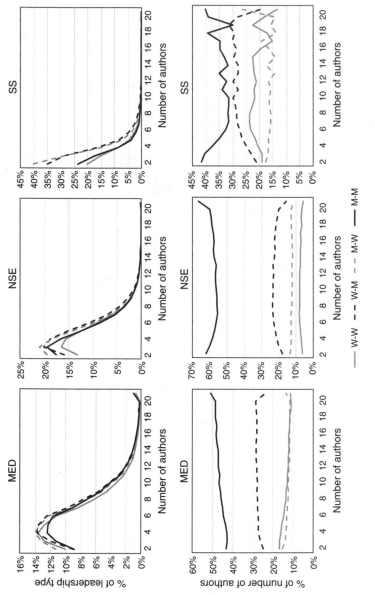

FIGURE 2.3. Distribution of the percentage of papers as a function of the gender of first and last authors, 2008–2020. Top row of panels: Percentage compiled within the collaboration type (x axis); sum of the line equals 100% (minus single-authored papers). Bottom row of panels: Percentage compiled within a number of authors' class (y axis); sum of a given number of authors' class equals 100%. MED: medicine; NSE: natural sciences and engineering; SS: social sciences; W-W: women first and last authors; W-M: woman first author and man last author; M-W: man first author and woman last author; M-M: men first and last authors. Only includes papers that have more than one author. Data from Web of Science core collection.

Gender Homophily in Dominant Authorship Positions

One of the reasons given to explain the lower proportion of women pursuing scientific careers is the lack of women in leadership roles on scientific articles. Previous research on the topic has provided mixed results. While large-scale analyses in the life sciences have provided evidence of gender homophily in the choice of collaborators, other studies have shown that, while men were more likely to favor collaboration with other men, women were not likely to favor other women.[54] To assess homophily, we focus on the relationship between first and last authorships. As mentioned previously, last authors are typically senior team leaders, and first authors are more likely to be junior scholars. Figure 2.4 presents, by discipline, the percentage of articles with women first authors, as a function of the gender of the last authors. All disciplines combined, the difference is

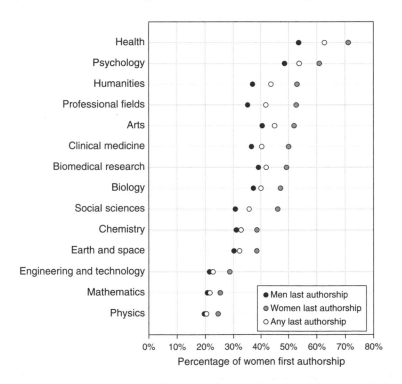

FIGURE 2.4. Percentage of women first authorship, as a function of the gender of the last author (women, men, any), by discipline, 2008–2020. Only includes papers that have between two and twenty authors. Data from Web of Science core collection.

striking: when women are in senior positions, they select women first authors in 46% of instances; when men are in senior positions, only 32% of first authors are women. This result is observed in each discipline: articles with a woman as last author are disproportionately likely to have a woman as first author. This is particularly the case in social sciences and humanities fields, where articles with a senior woman are 43%–50% more likely to have a woman first author than articles with a man as senior author. While it is not possible to determine which one of the two initiated the collaboration—first authors may choose to work with a given senior scholar, while last authors may offer a junior scholar a position in their lab—our results suggest strong gender homophily in collaboration. These roles are critical for retention in scientific careers, suggesting a clear policy rationale for advancing women and hiring senior women: representation in first authorship is critical to propel early-career researchers, and this is more likely to happen when there are a greater number of women in senior positions.[55]

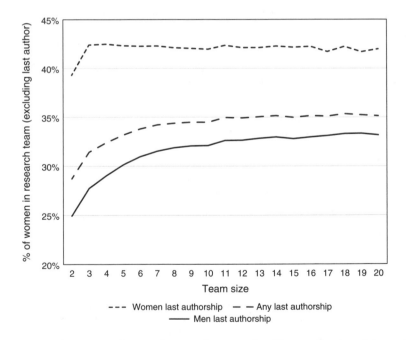

FIGURE 2.5. Percentage of women in research team authorship, as a function of the gender of the last author (women, men, any) and of team size (number of authors), 2008–2020. Only includes papers that have between two and twenty authors. Data from Web of Science core collection.

Team composition, however, is not limited to first and last authorship, as women are disproportionately found in middle authorships. To assess how leadership roles affect the gender composition of teams, we compiled the average percentage of women in authorship lists, as a function of the gender of the last author (Figure 2.5). Results show that, for teams that have between three and twenty authors, the percentage of women on the byline is stable at 42% when the last author is a woman.[56] When the last author is a man, the percentage of women in the research team increases as a function of team size—from an average of 25% for papers with two authors to 33% for papers that have between eleven and twenty authors—but remains sizably smaller than when women are in leadership positions. This reinforces findings from earlier research, which suggested that men were more likely to have more men graduate students on their teams.[57] The implications for this are strong: an increase in the percentage of women in senior roles may lead to more diverse research teams and the provision of more important roles for early-career women in science.

Global Collaborations

Collaborations are increasingly international.[58] Although percentages of international collaboration vary drastically by country—smaller countries generally having higher international collaboration rates, and larger countries higher rates of domestic collaboration—there is a general trend toward collaborations that span a wide geographic distance.[59] Figure 2.6 presents the percentage of national collaboration (collaboration with other colleagues from the same country) and international collaboration (papers with at least two countries represented) by discipline and gender of last authors. It shows that, irrespective of the leadership position, women tend to collaborate more domestically and men, internationally. This holds regardless of author order and is invariant across disciplines and countries. Gender differences are particularly prevalent in psychology, professional fields, social sciences, and mathematics and less pronounced in arts, humanities, chemistry, and clinical medicine.

Generally, we observe a weak relationship between the percentage of international collaboration in a domain and the gendered difference in international collaboration: the higher the percentage of international collaboration, the more likely men are to be active in those collaborations

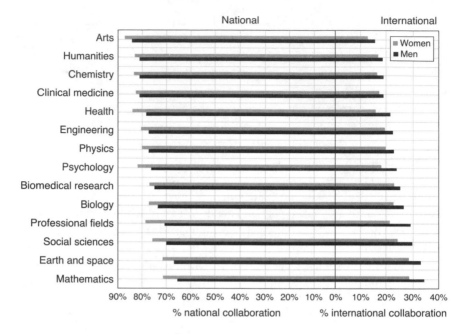

FIGURE 2.6. Percentage of collaborative papers that are in national or international collaboration, by discipline and gender of last author, 2008–2020. Only includes papers that have more than one author. Data from Web of Science core collection.

compared with women. Four specialties with high international collaboration rates follow a different logic: astronomy and astrophysics, earth and planetary science, meteorology and atmospheric science, and nuclear and particle physics. In those specialties, men and women exhibit similar rates of international collaboration. The commonality of these disciplines is the inherent anchoring of specific instrumentation or infrastructure, such as astronomical observatories and colliders. This suggests a selection effect, whereby women must comply with the habitus of the field, which demands international engagement.[60] We also observe this pattern of gender neutrality, though to a lesser extent, for subjects in the social sciences, such as area studies. Domains with large gender gaps may be those for which mobility is not a requirement to success though it certainly mediates performance, as we will explore in later chapters.

The lower participation of women in international collaborations is also visible for each country—irrespective of the profile of international collaboration for the specific country (Figure 2.7). For example, in the

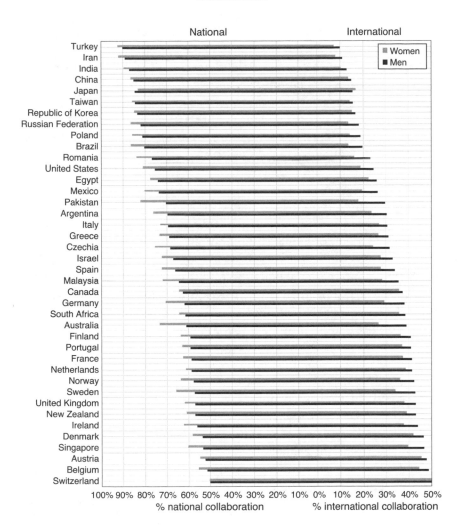

FIGURE 2.7. Percentage of collaborative papers that are in national or international collaboration, by country and gender of last author, 2008–2020. Only includes papers that have more than one author. For top forty countries with the highest numbers of papers. Data from Web of Science core collection.

United States—where a relatively low proportion of papers are the result of international collaboration—men's papers are 30% more likely to be the result of an international collaboration than those of their women colleagues. The only exception to this trend is Switzerland—the country where international collaboration is the highest among those presented in the figure—for which national and international collaboration are similar for both men and women.

Explanations for these gendered differences are complex, but they resonate with the findings on production, where we revealed that women's work is often targeted toward care disciplines.[61] Care disciplines (such as health and education) are inherently more nationally focused, given that they are contextualized within cultural and political systems. Another factor that undoubtedly contributes to these trends is women's caregiving responsibilities.[62] Although collaboration can increasingly be conducted through various online tools, several collaborative projects need in-person meetings, such as those associated with specific infrastructure. Given that caregiving can create burdens to travel, this may reduce women's participation in international collaboration.

Despite these differences, both men and women are increasing their percentage of international collaboration: from 19% in 2008 to 26% in 2020 for men, and from 16% to 22% over the same period for women. While the relative growth of women's international collaboration activities was higher than that of men, there remains a persistent gap in absolute percentages—6%—over the last thirteen years, which suggests that women are not bridging the gap with men's international collaboration activities. Scholars have suggested that the shock of the COVID-19 pandemic could reduce these inequities, given that travel was closed for men and women alike.[63] However, no such effect was observed for 2020. There could be several explanations for this. One could simply be a function of time: it takes time for changes in scholarship to be evident in the data, and the preexisting collaborations and their products have not yet begun to decay. It may also be a function of the lower rates of scholarly production for women during the pandemic.[64] It remains to be seen how the scholarly communication ecosystem will rebuild postpandemic and what the consequences will be for equity.

Collaboration across Sectors

Many sectors contribute to the research enterprise—universities, industries, and governments—each of which brings its own logics and institutional culture.[65] While the large majority of basic research, as measured by research articles, is performed in a university setting, most applied research, as measured by patents, is industry driven. This trend is growing:

while less than 70% of all research articles originated from universities in 1980, this percentage increased to almost 80% in 2014. Industry articles have been in a relative decline—from 7% to 3% of the overall output. This is largely emblematic of incentive structures—wherein publication is required for advancement in universities but is often optional or prohibited in industry. Rather, industry incentivizes commercialization: industry increased its share of all patents held from about 70% in 1976 to almost 90% in 2014. Concomitant with their decline in basic research activities, industries have been increasing dependence on universities for their basic research activities: more than 75% of recent papers with industrial affiliations are the result of collaboration with universities.[66] The benefits of cross-sectoral collaboration have been much heralded; however, there are several tensions that arise in terms of incentive systems for authorship and issues of ownership and intellectual property.[67] While tensions between commercial interests and scientific authorship have been present for centuries, these are made increasingly complex by the growing rates of collaboration among industry and academic scientists and the plurality of forms of funding for academic science.[68]

Our data show that university-industry relationships vary by gender (Figure 2.8). Previous research has shown that women were less likely to

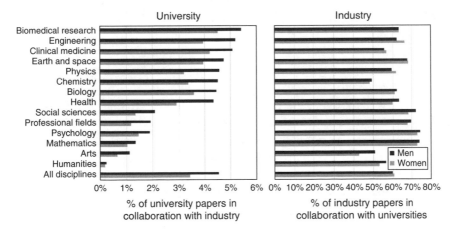

FIGURE 2.8. Percentage of university-led papers written in collaboration with industry (left), and of industry-led papers in collaboration with university (right), by gender and discipline, 2008–2020. Data from Web of Science core collection. Classifications of institutions into sectors were performed using the classifications developed in V. Larivière et al., "Vanishing Industries and the Rising Monopoly of Universities in Published Research," *PLOS ONE* 13, no. 8 (2018): e0202120.

collaborate with industrial partners—be it in terms of collaborative research projects or consulting agreements.[69] Our data reinforce these findings within the academic space: in every discipline, academic men are more likely to collaborate with industrial partners than women. However, we find a different trend for industrial researchers (that is, those authors with affiliations with an industry organization): in fields with low percentages of women researchers overall (such as engineering and physics), women in industry are more likely than men to collaborate with universities. These results may reflect a more hospitable climate in academe than in industry for publishing women. This reinforces studies of women in patenting, which shows that women are more likely to patent when in a university setting than an industrial setting.[70] Overall, this suggests different access to cross-sectoral collaborative networks between men and women, with implications for the receipt of capital and access to resources, particularly for those fields with institutionalized and incentivized university-industry relationships.

What Do the Authors Say?

The composition of scientific teams is not value neutral: there are many decisions around who becomes an author and the order in which they are placed. These decisions are not always amicable or unanimous. To understand the ways in which these decisions are made and disagreements that arise, we conducted a large-scale survey of contemporary authors. The results, drawn from over 5,000 respondents, demonstrated that more than half had encountered disagreements in either author naming or ordering.[71] Authorship disagreements are common in science, but women are more likely to experience disagreements and experience them more often.[72] This is particularly prevalent in the natural sciences and engineering, where there are significantly fewer women.

Authorship allocation is a deliberative process, but there are gendered differences in how, when, and by whom authorship is discussed. Scientists who discuss authorship have lower frequencies of authorship disputes than those who do not, regardless of gender. When authorship discussions are held, they are often led by or unilaterally determined by the principal investigator. In fact, the most common method is for the principal investigator to determine who will be an author and where they will appear, after consultation with the main collaborators. It is also commonplace for the principal investigator to make these determinations alone, particularly if the principal investigator is a man.

The disproportionate number of men in dominant author positions has a strong influence on authorship decisions. Men are more likely than women to determine authorship unilaterally at the end of a project. These practices may cause men to be more prone to the *recency effect,* providing stronger authorship positions to those who conducted tasks near the end of the project. Without consultation with the full team, they also may be unaware of all labor contributions, particularly of junior team members in a strongly hierarchical team. Women, on the other hand, are more likely to determine authorship at the beginning of the project, including the perspective of coauthors. The inclusion of multiple voices may lead to more equitable outcomes: discussion at inception tends to lead to lower rates of authorship disputes and more clarity among team members about what constitutes authorship.[73] It may, however, also prioritize the *primacy effect* in valuing contributions.

Both men and women report that the most probable cause of authorship disputes is "different ways of valuing or measuring the importance of the contribution," with women placing a greater emphasis on this than men. Women also noted "differing disciplinary practices" and "differencing ethics" as the cause for disputes, across all main disciplines. The different values are apparent when men and women are asked to attribute the value of certain tasks to research. In the natural sciences and engineering, women are more likely to value all tasks as more important than men, with the only insignificant variable being "technical work." Technical work and data collection were the only contributions in the medical sciences where men placed a higher value on the work. Men also rated technical work as having higher importance in the social sciences. As we will see in Chapter 3, this is also the type of task more likely to be associated with women. This suggests an important element in the self-(de)valuation of contributions by women.

There are consequences to authorship disagreements.[74] The most common outcome is limiting future collaborations. As we have shown in this chapter, collaboration is an essential component to contemporary academic work. Therefore, limitations in collaboration can have profound consequences for women's advancement in science. Women are also more likely to report hostility and sabotage around authorship issues. This, coupled with the high rates of other forms of harassment, contributes to negative work environments for women. Taken together, these results present a bleak picture: women are less likely to be in the conversation for authorship, are more likely to experience disputes, and are more likely to have negative consequences because of these disputes.

When asked explicitly, women are more likely than men to report that their colleagues are not distributing authorship in a fair manner. They report concerns about authorship distribution and an inability to comfortably discuss authorship. It is not surprising, given this evidence, that women report receiving less credit than they deserve. Inversely, men report receiving more credit than deserved.[75] This systemic devaluation of women's work in science creates cumulative disadvantage in scientific careers.

The Marriage of Matthew and Matilda

In 1977, sociologist Harriet Zuckerman published her tour de force, *Scientific Elite: Nobel Laureates in the United States*.[76] This book has been touted as a canonical work in understanding social stratification in science and includes comprehensive data—both sociodemographic and scientific—on Nobel laureates from 1901 to 1972.[77] Included were in-depth interviews with forty-one US laureates—all men. In a 1988 article, sociologist (and subsequently Zuckerman's husband) Robert K. Merton describes the interviews by Zuckerman as the source of his idea for a theory of cumulative advantage in science—that is, "that eminent scientists get disproportionately great credit for their contributions to science while relatively unknown ones tend to get disproportionately little for their occasionally comparable contributions." This theory, which Merton termed the Matthew effect, has had widespread influence on science studies. However, the foundational role of Zuckerman's data in the theory has received little attention. It is only through a footnote in a subsequent article by Merton, referencing a reprint, that we find this acknowledgment: "It is now [1973] belatedly evident to me that I drew upon the interview and other materials of the Zuckerman study to such an extent that, clearly, the paper should have been under joint authorship." Merton continues, "A sufficient sense of distributive and commutative justice requires one to recognize, however belatedly, that to write a scientific or scholarly paper is not necessarily sufficient grounds for designating oneself as its sole author."[78]

Merton's single-authored paper on the Matthew effect has received—to date—nearly as many citations as the collected published works of

Zuckerman. Zuckerman follows a long line of women whose visibility was diminished due to both informal and formal collaborations with their partners.[79] The irony, of course, is that this overshadowing was in the coining of a term to refer to this precise phenomenon: when the dominant actors receive disproportionate credit that is continually reinforced. Issues in authorship attribution, however, extend beyond the confines of domestic partnerships. Our results suggest that both men and women acknowledge inequities in the distribution of authorship along gendered lines.

The lack of women in senior roles has strong implications for the lack of women overall: nearly half of papers with a senior woman author (last author) have a woman as first author, whereas only a third of papers with men in senior positions have a woman in this dominant authorship position. These patterns of homophily extend to other authorship positions: more than 40% of all authors in a woman-led team are women, whereas the total proportion of women authors varies from one-quarter to one-third for men-led papers. The common placement is for women to be in middle or first authorship, which aligns with their unequal representation in senior ranks.[80] As team size increases, so too does the representation of men in prestigious first and last authorships. Taken together, these results suggest a clear policy implication: women in senior roles are more likely to incorporate women into science and provide them with leadership roles on papers. Therefore, support for senior women may also serve to reduce attrition for junior women.

This recommendation may be at odds with some of the literature on mentoring. Mentoring has been shown to be highly linked to career success in academe, particularly for women.[81] However, research on the role of same-gender dyadic relationships in mentorship is sparse and conflicted, and it often tends to conflate advising, mentorship, and role modeling (the last of which is more likely in same-gender mentorship relations).[82] Generally, the literature tends to reinforce the benefits of women being paired with men for mentoring, likely due to men's privileged space in the academic network and their disproportionate access to resources.[83] However, a study of chemistry doctorates found that having a same-gender adviser was linked to higher productivity and increased likelihood of academic placement for women advisees.[84] Other studies have reinforced the increased productivity in same-gender mentoring dyads and have suggested that women have a higher probability of graduating and a reduced time to graduation when they have women advisers.[85] The studies, however, remain highly limited and localized, calling for a more in-depth understanding of the role of same-gender advising and mentoring in doctoral education. Providing a larger diversity of scientific role models will be critical in determining the attrition and retention of a diverse scientific workforce.

Access to the broader scientific network is essential for career trajectories. However, we find that women are disadvantaged along several axes. They are more likely to work in collaboration, but the teams they lead tend to be smaller. Relative to men, women's collaboration networks are more domestic than international, and that is observed in all fields. Furthermore, women are less likely than men to collaborate with industrial partners. The main form of retaliation in authorship disputes— which women are more likely to experience and more often—is in limiting collaboration. All of this is more likely to exclude women from the broader scientific community, which limits their access to resources and knowledge.

To mitigate disparities and reduce the frequency of disputes, several senior researchers have adopted authorship guideline statements for their own research teams. One notable example is that of neuroscientist Stephen Kosslyn, who developed a point system for each type of labor contribution.[86] Other researchers tend to have more narrative guidelines, emulating institutional wording, such as that provided by the Office of the Provost at Yale University.[87] Guidelines such as these are increasing as institutions develop greater infrastructure around research integrity and conflict resolution for collaborative research.

The increasing ambiguity in authorship illuminates a central problem with contemporary collaborative work: inferring contribution from a byline of several authors leads to inaccuracies and misrepresentations. Furthermore, the emphasis on the byline itself is embedded in a broader conversation about goal displacement in science. It has been argued that the present scholarly communication system rewards a "taste for publication" over a "taste for science."[88] Deeper analysis of authorship and collaboration provides an evidence base from which to construct possible interventions that can incentivize behavior more aligned with the values and principles of scientific work. However, it may be that authorship itself is a flawed indicator for determining and signaling contributions. Therefore, in Chapter 3, we explore the issue of contributorship, which begins to unravel the opacity of collaborative authorship.

Chapter 3

Contributorship

E dward Pickering became director of the Harvard College Observatory in 1877. As the apocryphal story goes, some years later he became so frustrated with the inefficiency of his assistant (a man) that he declared his housecleaner would make a better technician. He promptly hired said maid, Williamina Paton Fleming (1857–1911), a twenty-four-year-old Scottish immigrant and single mother, as his first "computer."[1] She continued to work at the Observatory for the next thirty years, becoming one of the most well-known woman astronomers of her generation and setting the stage for the entrance of many more "computers" into astronomy. Initially, this growth was predominantly under her charge: between 1885 and 1900 she hired twenty women assistants, many of whom made significant contributions in the history of astrophysics and were eventually entered into the annals of the *American Men of Science*.

The success of hiring women at the Harvard College Observatory was emulated by other observatories around the world, leading to a new type of "women's work in astronomy." Such was the title of Fleming's address at the 1893 World's Fair in Chicago, where she sought to extol the virtues of computing work. In the speech, Fleming follows the rhetoric of gender stereotyping in computing, stating, "While we cannot maintain that in everything women is man's equal, yet in many things her patience, perseverance, and method made her his superior."[2] These unique feminine attributes for computing work were reinforced in the ways that others spoke about Fleming (and the entire cast of "Pickering's Women"). As a

65

journalist for the *Examiner* wrote, "The amount of labor devoted by Mrs. Fleming to the patient study, under the microscope, of thousands upon thousands of photographs of the spectra of stars would have daunted most men and would have been utterly beyond their capacity, for it require[s] a combination of patience and persistence, and faith, and minute accuracy which is, perhaps, rather a feminine than a masculine characteristic."[3]

That their talents were unique did not lead to equal pay. When Fleming inquired about a raise after two decades of service, she was informed that she received "an excellent salary as women's salaries stand."[4] She increased her hours to try to obtain a raise, but it was to no avail. She began to falter emotionally and physically, and by 1911, she had worked herself to death. In her obituary, Pickering commented that Fleming "formed a striking example of a woman who attained success in the higher paths of science without in any way losing the gifts of charm so characteristics of her sex."[5]

Woman assistants were considered essential, but their work as "computers" rather than astronomers classed them as "doers, not thinkers; workers, not scholars; amateurs, not professionals." This led to depressed salaries and fewer options for career advancement: "The field of astronomy had opened doors for women, but only to allow them to work in limited capacities. In the eyes of peers, computers were not transgressing but rather transferring domesticity to the observatory so that men might freely conduct the interpretive work of real science."[6]

The aura of domesticity was frequently applied to another woman computer, Annie Jump Cannon (1863–1941). Cannon was born in 1863 in Delaware and studied at Wellesley College, taking classes at Radcliffe College, and was hired at Harvard College Observatory as an assistant in 1896, under the mentorship of Fleming. She developed a widely adopted stellar classification and was recognized for the publication of catalogs of stellar spectra. In 1931, she became the first woman to win the Henry Draper Medal for research in astronomical physics. The journalism surrounding her award is a rich example of the perception of her contribution. A journalist wrote, "She would have been a first-rate housewife" but instead "took up light housekeeping among the stars." The piece continued, "'Oh, those untidy men folks,' we can hear Miss Cannon say as she took up astronomy. 'Let's get some order in this kitchen, I mean heaven.' So, she made her life work the cataloging of the stars. Hundreds of thousands of them she 'dusted off,' as it were, and put back into their right places . . . Housewives may be a little weak on astronomical physics. But they will understand just how Miss Cannon felt. Those heavens simply HAD to be tidied up."[7]

The early twentieth-century society could not comfortably envision a woman as an astrophysicist; the narrative, therefore, had to use domestic terms, to further codify the labor as women's work. Cannon herself cultivated this domestic persona, making herself unthreatening to the men around her. She functioned as a hostess to visitors and mother figure to other women in the Observatory. Like her predecessor Fleming, she also reinforced the unique talents of women, noting that women have the eyes and hands for this type of work, but that men "held the lead" in "interpretive" work. Despite her own devaluation, her contributions were acknowledged through membership in the Royal Astronomical Society, a number of honorary degrees, and the Ellen Richards Prize of the Association to Aid Scientific Research by Women.[8]

Cannon donated some of this prize money to establish an award with the American Astronomical Society for distinguished contributions of women scientists. The first prize, in 1934, was awarded to Cecilia Payne-Gaposchkin (1900–1979).[9] Born in England and educated at Cambridge University, Payne-Gaposchkin joined the Harvard College Observatory in 1923.[10] Harlow Shapley had replaced Pickering as director, and the old guard of women computers was slowly disappearing. Shapley encouraged Payne-Gaposchkin to write a dissertation, and she received the first doctoral degree in astronomy from the all-women Radcliffe College in 1925.[11] She was hired as an assistant at the Observatory, and although she made twice the salary of women computers, she still made less than men lecturers. In a letter detailing her consternation with the compensation and recognition she received in this role, she noted with great insult that the line item for her salary was not under personnel, but that she was paid as "equipment."[12] This is the most extreme manifestation of the instrumental status of women's work in astronomy in the early twentieth century. That is, women's contributions conferred on them a technical status as resources, rather than authors and scholars. Nevertheless, Payne-Gaposchkin persisted, and in 1956 she became Harvard's first woman astronomy professor and, subsequently, the first chairwoman of Harvard's astronomy department.[13]

From Collaboration to Contributorship

Production is an indicator of how much is produced, collaboration answers with whom, but contributorship tells us in what ways. Each of these indicators is uniquely and historically gendered. The increased demand for scientific technicians in the twentieth century had several consequences for science, including the feminization of certain types of labor within heavily men-dominated disciplines and an increase in the size of scientific teams. These concomitant changes led to broader concerns in accounting for scientific labor. Authorship was the primary means for accumulating capital in science—whether for hiring, awards, or priority disputes—and for assigning responsibility.[14] The dramatic changes to the scientific system wrought by increasing collaboration evoked serious consternation among members of the scientific community: by the mid-twentieth century, scholars began to express concerns that changes in the number of authors were the result of not merely a change in production but rather a change in fundamentally *accounting* for scientific labor. As noted by sociologist Harriet Zuckerman (1968),

> In science, evaluation is based on the extent and quality of scientific contributions as appraised by significant others in the field. This is nothing new. What is new is the profound change in the social organization of scientific work which, in many cases, reduces the visibility of role-performance of individual investigators. In the past, there were, as now, discrepant judgments about the quality of a scientist's work, but rarely was there any ambiguity about who had done what. A scientist may have had technical assistants working with him, but he was unambiguously *the* investigator, not merely the "principal investigator." When he put his findings into print, he was, with few exceptions, the sole author. There could be no doubt whose scientific performance was being assessed.[15]

These doubts, however, became more pronounced with the increased number of coauthors on a given publication. Author order was adopted by many disciplines as an implicit signal of the contribution of authors to a given work—either in descending order of contribution or using dominant first, last, and corresponding author positions. As Zuckerman explained, although it may seem "trivial or humiliating" for scientists to be concerned with author order, "name ordering is an adaptive device that facilitates the allocation of responsibility and credit among co-workers in otherwise ambiguous situations induced by the new structures of scientific research. It is designed to have the reward system operate with a degree of equity and adequacy."[16] This system worked relatively well for a few disciplines with limited numbers of coauthors. However, the system of implicitly weighting credit and contribution was challenged by increasing interdisciplinarity—which brought differences in authorship norms in conflict with one another—and the striking rise in the mean number of coauthors that occurred in the decades following.[17]

In the twilight of the twentieth century, new voices began to challenge authorship as the primary accounting mechanism for scientific labor. A group of medical scholars, guided by Drummond Rennie (nephrologist and deputy editor of the *Journal of the American Medical Association*), led the charge for what they called a "radical conceptual and systematic change" that would eliminate the "outmoded notion of an author in favor of the more useful and realistic one of contributor." Rennie and colleagues argued that the "the expansion in numbers of authors per article has tended to dilute accountability, while scarcely seeming to diminish credit." The case was made plainly: "Credit and accountability cannot be assessed unless the contributions of those named as authors are disclosed to the readers."[18] Their proposed model, which they called *contributorship*, required that a collaborative publication identifies not only who contributed to the work (as implied in authorship) but also what explicit contribution they made that warranted authorship.

The idea of contributorship found immediate favor among many leading medical journals, such as the *British Medical Journal* and the *Lancet*, arguably due to the importance not only of *credit* but of *responsibility* in medical research.[19] Provocatively written editorials from these editors heralded the death of authorship and called on the community to abandon traditional authorship practices.[20] Despite this early activism, contributorship was largely adopted as supplement to, rather than a replacement for, authorship. Authorship remained the dominant form of credit attribution, reinforced by the bibliographic reference and the formatting of academic curricula vitae. Contributorship statements, for their part, were largely buried in PDFs, as part of the acknowledgments or an explicit contributorship section, and therefore not readily discoverable for large-scale analyses.

One publisher served as the exception to this rule. From its inception in 2003, the Public Library of Science (PLOS) systematically requested that authors provide contributorship information for published manuscripts. These data were then published online with other corresponding metadata. The online platform of PLOS, the quality of the metadata surrounding the documents, and the availability of application program interfaces made it possible to download and parse full contributorship data. In the early years, author contributions were not standardized and included an array of free text and idiosyncratic contributorship statements.[21] By 2016, nearly all articles solidified into five main categories: analyzed the data, conceived or designed the experiments, performed the experiments, and wrote the paper. About three-quarters of papers indicated an individual who had contributed materials or resources. There were 20,000 other types of contributions that were listed, with about one-fifth of articles containing at least one of these. These nonclassified contributions demonstrate the extremely long tail of contributorship: there are core contributions that appear in nearly every scientific paper, but there are several thousand other types of scientific tasks across fields.[22] Finding a complete taxonomy that would incorporate all these tasks, and be functional across disciplines, remains a clear challenge, but one that is seeing renewed energy with the development and adoption of the CRediT taxonomy (as we will discuss later in this chapter).

Who Does What?

Contributorship data provide an invaluable lens to answer questions about gender disparities in the production of science. Given that much scientific work is produced in teams, contributorship provides a unique opportunity to examine—across disciplines and journals—the ways in which labor is distributed and enacted. Furthermore, given that dominant authorship roles are still used both implicitly and explicitly in research evaluation, it is important to understand the labor associated with these authorship positions and potential differences between genders. Analyses of contributorship provide insights into issues related to career progression and, thereby, have strong implications for changing the ways in which team science is conducted and how it can lead to more equitable outcomes. Contributorship research builds on the work of science studies scholars who argue for a move from focusing on the "content of scientific knowledge" toward "an ecology of the conditions of its production": a distinction framed by sociologist Andrew Pickering as "science as knowledge" versus "science as practice."[23]

In our early analysis of more than 600,000 contributorship statements across nearly 90,000 articles published by PLOS between 2003 and 2013, we were able to provide an initial understanding of the link between authorship and labor roles.[24] One of the most illuminating results of this analysis was the bifurcation of conceptual and technical work within research: that is, that those who are performing the actual experimentation are often isolated from the design and writing of the paper. We also found several invariant relationships between dominant authorship roles and contributorship tasks: first authors performed the bulk of contributions, followed by last authors, who were also the most likely to be corresponding author. Middle authors were associated with fewer tasks, regardless of discipline.[25]

Contributions are not independent from the sociodemographic characteristics of authors, such as demographic and academic age, gender, and organizational status (for example, rank).[26] Younger scholars tend to be in first or middle authorships and are more likely to be associated with experimentation as well as the spectrum of writing and analysis roles associated with first authors. Older and higher-status authors are more likely to be found in last author positions and tasked with contributing resources and conceptualizing the work. There are significant gender disparities in contemporary labor roles in science, with women more likely to earn authorship from performing experimental work, while men are typically associated with conceptual work and the contribution of resources.[27] Some may argue that this is merely an artifact of the leadership roles in science and will dissipate when there are more women as heads of labs. However, the pattern held true even when controlling for academic age: women are as likely to perform experiments as men who are ten years younger. Moreover, such disparities persist when controlling for corresponding author: when women are in this dominant position, other women on the team are more likely than men to contribute to all tasks, with the exception of contributing resources.[28] For men-led teams, men tend to contribute to the majority of tasks, with the exception of performing experimentation. This reinforces and extends our finding on women's leadership in collaboratively authored articles (Chapter 2).

Collaboration and team size affect the division of scientific labor. Previous evidence shows that the configuration of teams, and the division of labor within them, is a function of the gender of those who are in leadership roles. For example, the number of tasks performed by individual researchers in women-led teams does not diminish as quickly with the addition of new authors. That is, despite an increase in team size, a large proportion of women continue to contribute in all ways, particularly in experimentation. Men on men-led teams, however, follow the adage "Many hands make light work": the proportion of men doing any single task decreases steadily as the first ten members of the team are added to the work.[29] Women on women-led teams continue to do the majority of the experimentation, even when new team members are added. Our early work suggested strong differences in how men and women are employed in scientific work and gendered differences in the distribution of labor in scientific teams. However, the work was limited by the coarse classification system and the idiosyncratic practices of naming contributors.

Adoption of CRediT

In 2016, PLOS adopted a formal taxonomy for contributorship.[30] The origins of the taxonomy can be traced back to a group of journal editors who met in 2012 under the institutional umbrellas of Harvard University and the Wellcome Trust to develop a contributor role taxonomy, expanding on previous classifications employed by PLOS and other platforms.[31] A fourteen-role taxonomy was constructed and assessed via a survey sent to 1,200 corresponding authors from several publishers. Authors were asked to apply the taxonomy to their own work and evaluate it.[32] Of the 230 respondents, most had favorable feedback. Following this initial test, the organizers partnered with two standards organizations—the Consortia Advancing Standards in Research Administration Information and the National Information Standards Organization—to continue to develop the taxonomy.[33] In 2014, a seventeen-person working group of publishers, funders, and academics began work on what was then titled Project CRediT (Contributor Roles Taxonomy) and released a taxonomy for public comment. By 2015, ORCID had begun working on badges to accompany Project CRediT. With minor clarifications, the taxonomy became stable and was adopted by PLOS in 2016. More than 120 journals—including *Cell, eLife,* and *F1000*—have since adopted the taxonomy, and many of them have argued for widespread adoption to create a standard taxonomy for contributorship.[34]

CRediT provides increased transparency for accounting for gender divisions in scientific production. The fourteen-part taxonomy separates many tasks that were aggregated in the original PLOS five-part categorization. For example, the contribution of writing is divided into two tasks: "writing—review and editing" and "writing—original draft." This division acknowledges that drafting and editing are fundamentally different tasks. While all authors may be able to claim review and editing, far fewer are responsible for penning the original text. The refined CRediT taxonomy also incorporates many contributions that were not explicit in previous classifications, such as supervision, funding acquisition, and project administration—which were generally implicitly covered in the resource category—as well as visualization, software, and data curation, which are arguably components of analysis and data collection.

These distinctions have strong implications for understanding gendered labor divisions in science. Observing the previous categorization scheme, it appeared that men dominated the writing of the manuscript.[35] However, this early categorization did not distinguish the various stages of writing and included any contribution to the writing of the paper, from the writing of the complete first draft of the manuscript to light editing. Through analysis of the more than 30,770 papers published in PLOS journals between June 2017 and December 2018 using the CRediT taxonomy, a new picture emerges: men are 9% more likely to review and edit the manuscript—a task that is also more likely to be performed by any author—while women are 6% more likely to have written the original draft.[36] This demonstrates that the original findings were skewed by the ubiquity of the editing portion of writing. Once the original drafting is separated from making edits to the text, we clearly observe that women are more likely to perform most of the writing associated with scholarly articles.

Results from articles utilizing the CRediT taxonomy also confirm some of the findings obtained using the original PLOS taxonomy, especially in terms of the gender divide between conceptual and empirical work (Figure 3.1). Women are clearly associated with empirical work: they are 18% more likely than men to contribute to the investigation and 16% more likely to have contributed to data curation. Men, on the other hand, are overrepresented in tasks that are either more conceptual or associated with seniority. For instance, men are 40% more likely than women to have contributed to funding acquisition and 36% more likely to be associated with supervision. Men are also sizably more associated with contribution

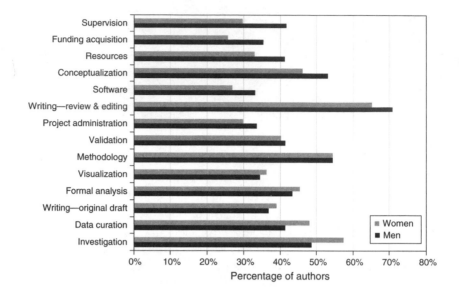

FIGURE 3.1. Percentage of authors who have performed a specific CRediT contribution, by gender (30,054 papers published in PLOS journals between June 15, 2017, and December 31, 2018).

to resources (24%), software (23%), conceptualization (15%), and project administration (12%).

In the largest PLOS journal, the multidisciplinary *PLOS ONE,* women are disproportionately associated with investigation, formal analysis, data curation, and visualization (Table 3.1). Similar trends can be observed across all PLOS journals, with small differences. For example, women contribute more to "methodology" than men in *PLOS Neglected Tropical Diseases, PLOS Pathogens,* and *PLOS Medicine.* In *PLOS Medicine,* men have a larger share of "investigation" compared with the other journals, where women hold a larger share of this contribution type. Such differences bring into question the idiosyncratic interpretation of such terms. For example, in medicine, it is possible that "investigation" is not associated with instrumental aspects of investigation but rather used to reflect the notion of a principal investigator—that is, the one who has received funding for the project. *PLOS Genetics* follows the generic PLOS patterns, but with greater disparities in the differences between men supervisory roles and women investigation roles. The strongest differences, from a gender perspective, can be found in *PLOS Computational Biology,*

TABLE 3.1. Difference in the percentage of authors who have performed a specific CRediT contribution, by gender and journal (30,054 papers published in PLOS journals between June 15, 2017, and December 31, 2018). Positive (darker cells) percentages mean that men are more likely to perform the task, and negative (lighter cells) percentages mean that women are more likely to perform the task.

Contribution	PLOS Comp. Biol.	PLOS Gen.	PLOS Path.	PLOS ONE	PLOS Med.	PLOS Neg. Trop. Dis.
Funding acquisition	10	15	15	9	7	9
Supervision	11	15	15	12	4	9
Resources	6	11	15	8	6	5
Conceptualization	13	8	9	7	8	4
Software	13	9	9	6	6	4
Writing—review and editing	6	8	8	5	1	4
Project administration	5	10	9	3	−5	4
Methodology	10	1	−3	0	0	−3
Formal analysis	10	1	−2	−2	−1	0
Writing—original draft	7	3	2	−3	−1	−4
Validation	4	−3	−1	1	0	−2
Visualization	7	−1	−4	−2	0	−3
Data curation	−5	−4	−5	−7	−2	−4
Investigation	−3	−15	−15	−8	−1	−9

where men dominated all categories save two: investigation and data curation. This suggests that computational biology reflects more of the gender dimensions of its computational father, rather than its biological mother, and harks back to the history of women computers.[37]

There are, as noted in previous work, strong variations by author order.[38] Last authors are more likely to perform tasks associated with seniority: reviewing the final draft, supervising the research, conceptualizing the research project, obtaining funding, doing project administration, and providing resources. First authors are more likely to be doing the bulk of instrumental tasks: curating and analyzing the data, designing the methodology, validating and visualizing the data, contributing to software, and writing the original draft. Middle authors contribute less, with strengths in investigation, data curation, and methodology.

Gender differences are also observed within these positions (Figure 3.2), as our results on collaboration suggested (Chapter 2). The percentage of men and women last authors performing any given task is similar, but women are proportionally more likely to be associated with all tasks except for funding acquisition and software. This suggests that, in senior roles, women contribute at a higher rate to tasks than senior author men. This could be interpreted in several ways: in a positive light, we can suggest that women are more engaged with the tasks and therefore more likely to identify misconduct and error. However, this mode of production is also less efficient, which could be an explanation for their lower productivity. Either way, the results suggest that senior women do more work per paper than their men colleagues. In first authorship roles, the trend is reversed—a higher proportion of men in that position perform each task, apart from investigation. This demonstrates that men are in stronger leadership positions during their junior years, with implications for career progression. This held true in a subsequent analysis, where we found that men were more likely to emerge within the "leader archetype" compared with women, who specialized early in their careers and were, consequently, more likely to experience attrition.[39]

Middle authors present some of the most interestingly gendered differences, with men receiving middle author credit for many of the same tasks as last author: supervision, conceptualization, and contribution of resources. Women in middle author positions, however, are likely to be doing the experimental and data work for the project. This has important implications for the findings on production and productivity discussed in earlier chapters. Our data suggest that senior men authors are brought onto bylines by duplicating last author roles (and possibly honorific authorship), whereas women receive middle authorship for instrumental work. These tasks differ dramatically in time investment, with women's work occupying much more time. Furthermore, by dividing conceptual and supervisory tasks, men can maximize efficiency in production. The contributions of women last authors suggest that senior women are not engaging in a similar division of labor. Furthermore, first author men seem to be offered more opportunities to take leadership roles on the team.

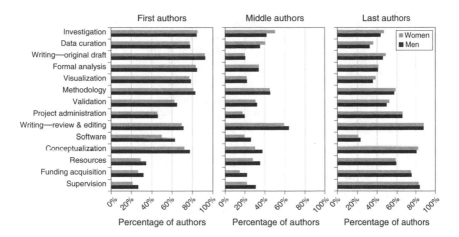

FIGURE 3.2. Percentage of authors who have performed a specific CRediT contribution, by gender and author order (30,054 papers published between June 15, 2017, and December 31, 2018).

These patterns of labor division serve to reify status hierarchies in academe, relegating women technicians to lower-status positions and reducing their likelihood of obtaining coveted first author positions, essential for careers in science.[40]

This raises the question: are men and women corresponding authors performing the same tasks on the manuscripts that they lead? Figure 3.3 shows that the gender differences in contributions observed for different author orders are also observed for corresponding authors, with women corresponding authors more likely to have contributed to empirical tasks, and men corresponding authors associated with conceptual contributions. This provides key insights into how scholars are utilized on a team and given credit for their work. What emerges from this analysis is a demonstration of a gender-balanced woman-led team in which the woman lead is embedded in the research process and provides nominal authorship credit to those who contribute resources. Men-led teams, on the other hand, incorporate junior scholars as middle authors who contribute writing, and women authors disproportionately contribute to intensive labor roles. Men as team leaders tend to be less involved in the tasks of the project, delegating these tasks to first authors.

CRediT contributions are typically assigned by the corresponding author. Given that the corresponding author is usually either first or last

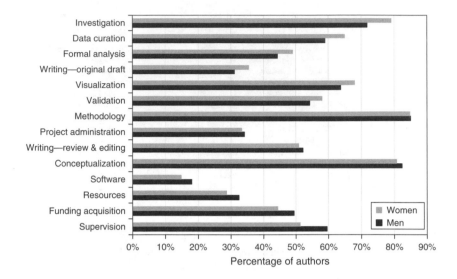

FIGURE 3.3. Percentage of women and men as corresponding authors who have performed a specific CRediT contribution, for women and men corresponding authors. (30,054 papers published between June 15, 2017, and December 31, 2018).

author, it is expected that there will be a strong trend of same-gender credit assignment, with the dominant gender assignment following the gender of the corresponding author. These trends are not necessarily a demonstration of bias, but a reflection of corresponding authors' self-crediting. To better understand how credit is allocated and labor divided in teams led by men and women, we removed corresponding authors from the author list and analyzed how labor is divided across the rest of the team (Figure 3.4). This analysis reveals that the gender gap in tasks performed is not affected by the gender of team leaders; tasks that are generally performed by women (investigation, data curation, and formal analysis) remain so irrespective of the gender of the corresponding author. The writing of the original draft is slightly more likely to be performed by women when the corresponding author is of the same gender. Similarly, tasks that are more often performed by men (supervision, funding acquisition, resources, software, conceptualization, review and editing, and project administration) are not affected by the gender of the corresponding author. The only observed shifts are in validation and methodology: they are slightly more likely to be performed by authors of the same gender as that of the corresponding author. Overall, this suggests that labor division is

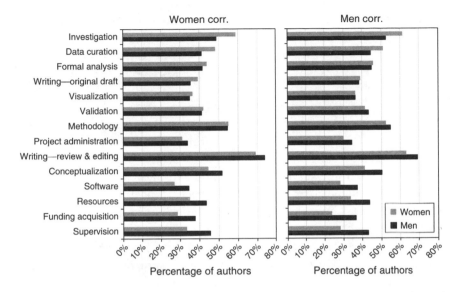

FIGURE 3.4. Percentage of authors (excluding corresponding authors) who have performed a specific CRediT contribution, by gender of corresponding author (30,054 papers published between June 15, 2017, and December 31, 2018).

not drastically affected by the gender of the corresponding author and that researchers of a given gender roughly perform the same tasks irrespective of the gender of team leaders.

The Author's Perspective

One of the limitations of the PLOS data is that it concentrates on the biological and medical sciences: more than 90% of papers published in the multidisciplinary journal *PLOS ONE* come from one of those disciplines.[41] Moreover, a strong tension emerges in the analysis of contributorship data between the actual contribution on teams and gendered differences in the allocation of credit. The mix of gender politics and academic hierarchies may compromise the accuracy of contributorship statements. These are public documents, and there may be pressures both to include an individual as an author and to overstate their contributions.

To better understand gender differences in contributions, we complemented the unobtrusive bibliometric analyses with an obtrusive survey that asked corresponding authors what they and their team members contributed to the research.[42] The survey respondents were drawn from the full population of corresponding authors on papers indexed in the Web of Science, and they cover all disciplines—except arts and humanities, which were excluded because of their low levels of collaboration and, therefore, of division of labor. We used the five PLOS tasks as a foundation but added an additional category—data collection—to the analysis. More than 12,000 authors responded to the survey, with a gender distribution that matches the general distribution of authorships (30% women).[43]

The results of the survey strongly validated the bibliometric data but also provided some complementary evidence. Contrary to what was observed for PLOS papers, women corresponding authors reported contributing to all tasks more than men corresponding authors, except for resource contributions (Figure 3.5). They were overwhelmingly responsible for data collection: a task often given to middle or first authors, rather than senior corresponding authors. This suggests that women principal investigators manage their teams differently from men and that they are more involved in the process of doing research. There were also gendered differences in the tasks men and women were likely to credit on a given paper. Men corresponding authors were likely to credit a larger percentage of authors with performing experimentation, doing data collection,

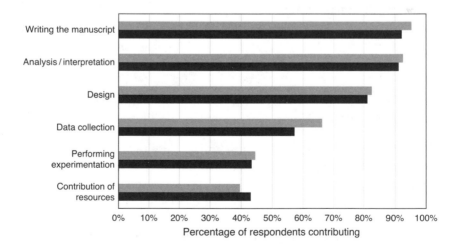

FIGURE 3.5. Proportion of survey respondents (corresponding authors) contributing to specific tasks, by gender ($N = 12{,}001$ corresponding authors of Web of Science–indexed papers published in 2016).

and contributing resources. Women, on the other hand, were more likely to credit individuals for writing the manuscript and contributing to research design.

However, there were very few differences in how these tasks were attributed to men or women coauthors, suggesting that differences in contributorship are the result of gendered differences in contribution, rather than in the allocation of credit. While there are no differences in the proportion of women and men authors who have analyzed or interpreted the data, it is clear that women researchers are more likely to contribute to data collection and perform experimentation—and the gap remains similar irrespective of the gender of the corresponding author (Figure 3.6). The only strong deviation was in the allocation of credit for writing: women corresponding authors were likely to report slightly lower contributions from men coauthors in writing. This difference may be explained by the nuances made evident in the CRediT taxonomy: women were more likely to contribute to writing the original draft, whereas men's contribution to writing was largely in reviewing and editing.

There is a strong relationship between author order and the distribution of labor, with first authors doing a disproportionate amount of work, and last authors holding another dominant place in author order (Figure 3.7). Women are more likely to collect data and perform experiments, while men are more likely to contribute resources and, to a lesser extent, research design. There are few differences in design, analysis, and

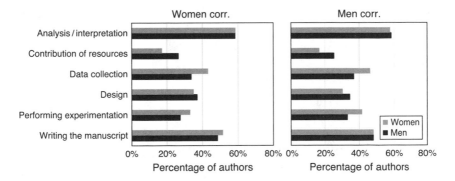

FIGURE 3.6. Percentage of authors (other than corresponding) who have performed a specific task, by gender, women corresponding authors (left panel) and men corresponding authors (right panel) ($N = 43,699$ contributions assigned by $11,538$ corresponding authors of Web of Science–indexed papers published in 2016).

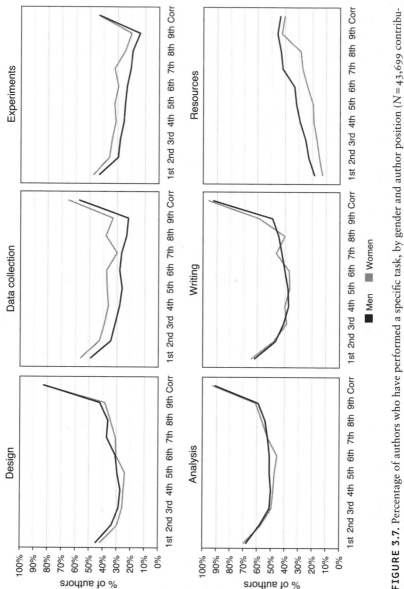

FIGURE 3.7. Percentage of authors who have performed a specific task, by gender and author position ($N = 43,699$ contributions assigned by 11,538 corresponding authors of Web of Science–indexed papers published in 2016).

writing, which suggests that differences observed earlier in those contributions are mostly a function of differences in author order.

There are, however, some nuances that can be observed, such as disciplinary differences in gendered divisions of labor in analysis and writing (Table 3.2). For example, writing tends to be more associated with men

TABLE 3.2. Difference between the percentage of men and women authors who have performed a specific contribution, by discipline ($N = 55,237$ contributions assigned by $11,538$ corresponding authors of Web of Science–indexed papers published in 2016). Negative percentages (lighter cells) mean that women are more likely to perform a contribution, and positive percentages (darker cells) mean that men are more likely to perform a contribution.

Discipline	Resources	Design	Analysis or interpretation	Writing	Experiments	Data collection
Biology ($n = 1,223$)	28	12	−1	4	−37	−16
Biomedical research ($n = 1,336$)	35	18	8	11	−24	−21
Chemistry ($n = 884$)	27	28	7	13	−25	−7
Clinical medicine ($n = 2,299$)	29	13	2	0	−14	−19
Earth and space ($n = 1,097$)	25	−1	−2	−7	−12	−7
Engineering ($n = 2,093$)	19	6	5	3	−15	−13
Health ($n = 380$)	16	0	0	−3	0	−23
Mathematics ($n = 388$)	−30	1	−6	−4	−18	−8
Physics ($n = 865$)	22	21	−1	5	−23	−21
Professional fields ($n = 538$)	31	6	−5	−9	−11	−23
Psychology ($n = 387$)	31	10	9	1	−5	−18
Social sciences ($n = 511$)	21	0	−8	−8	1	−23

in biology, biomedical research, chemistry, and physics; it is more likely to be performed by women in the professional fields, social sciences, and earth and space science. Analysis tends to be feminized in the social sciences and mathematics, whereas it is a masculine task in psychology and biomedical sciences. Health and the social sciences observe gender parity in design, and health is the most likely to achieve gender parity overall (though they still skew on the side of women for data collection and on the side of men for the contribution of resources).

In sum, the survey data are coherent with the previous analyses of gender differences in contribution, serving as a validation for both the gender assignment and the use of contributorship data to reflect actual contribution.[44] The survey results show high correlation between the algorithmically assigned gender and self-identified data from the survey (see Table A.1 in the appendix). Furthermore, the consistency of the contributorship data suggests that these metadata are appropriate for large-scale evaluation and reflect true labor contributions.

Sustained Gender Segregation in the Sciences

The gendered nature of tasks and the differential value attached have been persistent across the century. For example, in 1953, three articles were published sequentially in a single issue of *Nature,* each addressing discoveries related to the structure of DNA, but with differing emphases. The first article, written by James Watson and Francis Crick, was theoretical in nature, containing a "purely diagrammatic" figure and acknowledging that the ideas were "stimulated by a knowledge of the general nature of the unpublished experimental results and ideas of Dr. M.H.F. Wilkins, Dr. R.E. Franklin and their coworkers."[45] A second, data-driven article was published by Maurice Wilkins and two other colleagues. The third article, authored by chemist Rosalind Franklin and doctoral student Ray Gosling, was highly analytical, contained "Photo 51," and is credited for its precision and contribution to the discovery of DNA. It is well established that Franklin's data were instrumental for Crick's calculations, yet, in the history of this discovery, it is not Franklin's data but rather Crick's diagram that received credit.

Franklin died before the Nobel Prize was jointly awarded to Crick, Watson, and Wilkins in 1962. We are not in the position to argue whether her gender would have kept her from joining that lauded triad—replacing one of the men, as the prize is capped at three individuals—were it not for her untimely death. Rather, it bears questioning whether her *contribution* would have been recognized as sufficient.[46] This is an important element in the story: perhaps Franklin's contribution was overlooked not because she was a woman rather than a man but because her contribution was empirical rather than theoretical.

Our work suggests that certain contributions are more lucrative, from the perspective of academic capital, and these are more likely to be associated with men. Therefore, either the distribution of labor or the reward system of science must be reexamined. Research careers can take different forms: our previous work suggests three scientist archetypes, as a function of the contributions they perform on papers: leaders, specialists, and supporting scientists.[47] Research leaders are more likely to design and write papers, while specialists contribute to experiments and help with study design, analysis, and writing. Supporting authors also make important contributions to the research projects in which they participate. If we maintain the status quo, individuals who dedicate their work to experimentation and the construction of high-quality datasets will continue to remain undervalued and are more likely to leave the scientific system. Furthermore, the relegation of certain segments of the research workforce to limited roles and contributions diminishes their potential for creativity and innovation throughout the research process.

Women entered the twentieth-century scientific workforce as technicians—as specialists or supporting scientists. Nearly a century has passed, and they continue to carry this mantle, despite advances in leadership roles in science. The role of "women as doers" and "men as thinkers" holds constant in our data, with women more likely to garner authorship from experimental work and men more likely to be credited with conceptual work and the contribution of resources. The lab still reflects the same type of hierarchy observed more than a century ago: a Pickering-type scientist leading and funding the research, with women assistants doing the manual labor of the work and often the invisible conceptual work.

Even when controlling for seniority, these patterns persist. As senior authors, women remain more likely to do experimental work and credit other authors for the contribution of resources. There are also gendered distinctions within tasks: women are more likely to do the original drafting, whereas men are more likely to review and edit. There remains a gendered difference among senior authors, suggesting that disparities are perpetuated with certain roles developed in undergraduate, graduate, and early career years that persist throughout the academic lifecycle. Credit for leading discoveries is, therefore, largely given to men, who constitute a disproportionate number of lead authorships. While men account for 65% of first authorships, they represent about 75% of last authorships (as shown in Chapter 2). This suggests sustained gender segregation in research activity and reinforces earlier studies on gender and "scientific style" as adaptive behavior in collaboration.[48]

The growth in the size of research teams and the corresponding rise in acknowledgments of a diversity of contributions have increased the number of people who receive credit for scientific production, but all individuals do not contribute equally. First authors are the dominant contributors and those most likely to adhere to the International Committee of Medical Journal Editors' guidelines on contributing from conception through analysis and to writing.[49] Few of the other authors meet all these conditions. For example, in clinical medicine and biomedical research, nearly 30% of authors are only associated with a single task. This automatically places them in violation of the guidelines, as they cannot have contributed to both writing and another task. If we look across fields, there is no field in which 30% of authors do all five of the main tasks identified by PLOS: writing, design, experimentation, analysis, and contribution of resources. The most isolated task of these is experimentation. That is, those people who do experimentation are most likely to be associated with a single contribution. This is highly problematic. It means that those who undertake data collection and experimentation are the most divorced from other aspects—such as research design, analysis, and writing. This may explain the concurrent rise of both team science and fraud: when scientific tasks become increasingly complex and independent, there is an increased opportunity for malicious behavior to be hidden from other coauthors.[50] Middle authors are the most likely to do isolated tasks.[51] Last authors contribute substantially, but largely in conceptual tasks and in the contribution of resources.

Circumventing these patterns will require a radical reconceptualization of labor roles in science. This transformation necessitates action at all levels: from federal and institutional policies to behaviors within an individual lab. Leaders of research teams (principal investigators) are particularly responsible, as they explicitly distribute labor and implicitly enable gender segregation in research. Those mentoring junior scholars should take care to distribute tasks equally among members of the team, developing areas of deficiency and ensuring that men and women alike are given opportunities for leadership in the conception and design of research projects. Women who do not assert themselves in these roles must be given opportunities and mentoring to perform them. Furthermore, there should be more explicit acknowledgment of the value of certain labor types: for example, women are disproportionately more likely to do experimentation; however, as we discussed in earlier chapters, this is the labor role that they value the least.[52] It is essential that scientific labor is appropriately acknowledged and rewarded. As management theorist Stephen R. Barley and organizational ethnographer Beth A. Bechky observed three decades ago, technicians "play a critical role in the production of scientific knowledge"; however, they tend to experience "status inconsistencies."[53] Their early warning still warrants repetition: "Unless conceptions of work rooted in the industrial revolution are revised, any recognition of the horizontal and interdependent occupation division of labor that characterized the science labs we investigated is likely to be straightjacketed by continued replication of organizational practices born of a vertical division of labor."[54]

Data on contributorship make explicit what is hidden behind lab doors, and therefore, contributes to more transparency in the allocation of reward and responsibility in science.[55] However, current analyses on contributorship are limited by the few journals that both collect these data and make them readily available for analysis. Furthermore, only in a minority of papers (43%) are all authors involved in discussions of contributorship statements.[56] Mechanisms that allow authors to individually acknowledge their contributions may yield different results, particularly for authors in lower-status positions. Following the criteria for journal indicators suggested by science and technology studies scholar Paul Wouters and colleagues,[57] journals should move toward stronger infrastructure for collecting and making contributorship data available and ready for analysis.

Our results show that the largest gender gaps are found in three highly related tasks: funding acquisition, provision of resources, and supervision of the project, with men mainly dominating in these areas. This is one of the issues that cannot be readily addressed by a change in lab behaviors. Rather, it is an issue for intervention at federal and institutional levels—for example, at all stages of grant funding. There are several consequences that stem from the unequal distribution of resources in science. If women and men are given different levels of resources in start-up packages and internal and external grant competitions, it will affect their ability to contribute in certain ways to the production of science. Science is a complex system in which there are several interdependencies: contributorship is dependent on the inputs of funding and a robust scientific workforce; in turn, it contributes to production and the allocation of symbolic capital. We move, therefore, to looking at the gendered nature of the allocation of resources for science.

Chapter 4

Funding

In an interview with American news editor Marie Mattingly Meloney for a profile in the *Delineator,* Nobel laureate Marie Curie (1867–1934) detailed her frustrations with obtaining the necessary resources to conduct science.[1] The benevolence of Andrew Carnegie had provided the funds necessary to keep Curie's lab staffed after the death of her husband, physicist Pierre Curie. Her immediate concern, however, was her diminishing supply of radium, which was essential for her research. Despite having discovered the element, she only had a single gram in her laboratory. Meloney was shocked to hear this and asked how many grams Curie had personally. "I? Oh, I have none. It belongs to my laboratory." Meloney responded that "surely revenues from the patent of radium can pay for more." Curie corrected her: "There were no patents. We were working in the interest of science. Radium is an element. It belongs to all people."[2]

Unfortunately, not everyone agreed that elements were public goods. More than half of the radium in existence was produced by the Standard Chemical Company of Pittsburgh, which sold it at $120 per milligram. Curie needed at least another gram. Meloney, therefore, dedicated herself to a fund-raising campaign for Curie, promoting her as a saintly figure: a mother, a martyr, and a medic. Meloney could not sway the public on the rationale of basic science alone—dispassionate physics and chemistry would not appeal to her audience—rather, she needed to demonstrate the direct benefit to society. Meloney, therefore, framed the motivation as one that would certainly lead to a cure for cancer. In exchange for this assistance, Curie was used as a figure to advance the causes of several women's associations in the United States. She was not entirely comfortable with this arrangement but acquiesced to obtain the resources she needed to do her scientific work. The fund-raising was a success in the end: Curie received $100,000 worth of radium, and nearly $150,000 worth of other prizes and materials. In addition, she received a $50,000 advance for an autobiography. The cost to Curie was an extended tour in the United States, marketing both her research and herself. The scientist as popularizer was not a role that came easily to her. Public appeal, however, was necessary for her to obtain the resources she required to conduct science.

Meloney was not the only woman who worked to gather resources for other women scientists. Ida Henrietta Hyde (1857–1945) was an American physiologist born ten years before Curie. Hyde understood scarcity of resources: her mother worked for a cleaner and seamstress in Chicago until she was able to start her own business. Unfortunately, her home and mother's business were destroyed in the Great Fire of Chicago (1871), forcing Hyde to enter the workforce at an early age.[3] It was with Hyde's earnings that her younger brother was able to attend the University of Illinois. Her own schooling was more incidental: she began to study biology through a book she chanced upon at the store where she worked. She attended night classes at the Chicago Athenaeum until she could pass the entrance exams necessary to gain admittance to the University of Illinois. Unfortunately, her savings only supported her for a single year, after which she dropped out of college and became a public school teacher. Hyde was thirty-one when she was finally able to enroll at Cornell University. She earned a scholarship to Bryn Mawr, worked at Woods Hole Biological Laboratory, and then received a European Fellowship from the Association of Collegiate Alumnae (the precursor to the American Association of University Women). She went first to Strasbourg, where she petitioned to matriculate at the doctoral level—the first woman in Germany in the natural sciences and mathematics to do so. She withdrew her petition amid gender-related pressure and moved it to Heidelberg. After several obstacles, she received her doctoral degree in 1896.

In that same year, Hyde was invited to occupy the Heidelberg University "table" at the Stazione Zoologica (Naples Zoological Station). This station was founded in 1872 by Anton Dohrn, a German biologist from a wealthy family. Dohrn had used his own resources and solicited support from the German government as well as several prominent scientists (including Charles Darwin) to establish the station. The goal was to create collective spaces and resources for the international zoological community. The station operated on the "bench system," whereby "tables" were rented out to different organizations for an annual subscription of $500.[4] By Dohrn's death in 1909, more than 2,200 scientists from Europe and the United States had worked at the Stazione Zoologica, including Hyde.

Hyde's experience in Naples was transformative. At Heidelberg, her professor had disallowed her from attending lectures or laboratory demonstrations. At Stazione Zoologica, she was able to work alongside researchers in the international community. In her own words, "What a rare privilege it was to pursue studies in this highly endowed station! Through the investigators who came here from all quarters of the globe it offered a center for interchange that led to international understanding and enduring friendships . . . Grateful for the generous spirit that pervaded all departments of the Station and the valuable benefits offered to men and women alike, I resolved upon returning to the United States to do all in my power to enable eligible women scientists to avail themselves of the laboratory's unexcelled opportunities."[5]

She did this by convening a meeting of wealthy women and leading women educators who were sympathetic to the cause of women in science, a group she called the Naples Table Association for Promoting Scientific Research by Women. Their attempt to raise the table fee through individual donations was unsuccessful; therefore, they turned to colleges for support, targeting those with a high number of women matriculates. They were successful and their petition for a table was accepted by Dohrn, though not without some reluctance. As Dohrn wrote in response to the request for a table,

> Let me openly and sincerely confess, that it has taken long years to persuade or rather convince me that the modern movement in favour of women's emancipation is a sound one. In fact, I am only half open to believe in a successful end of it, and would be glad, if it went on in a more moderate degree, than usually proclaimed. But there is one part of it, for which I have not hesitated to feel and confess a strong sympathy, that is the throwing open to women the pursuit of science and the higher intellectual development. I do not only believe women capable of higher intellectual training, but think it will be of the utmost advantage to them and to mankind, when wives and mothers share in those accomplishments, which make a difference in the educated and non-educated intellect. To share the life of an intellectual husband, *to*

direct the nursery, and prepare the future generations are such tasks, that (require) *only the very best* instruction and mental education *will do for it—and for those who do not win in the lottery of life and cannot partake in one way or another these highest functions of a woman, the quiet pursuit of some intellectual career must be granted to atone for the loss of better chances.*[6]

The Naples Table Association for Promoting Scientific Research by Women was active until 1933. During this time, $17,000 in research grants and prizes was awarded and nearly forty women occupied the table in Naples.

Roger Arliner Young (1889–1964) was an American zoologist who would have benefited from these types of resources. She grew up impoverished and spent much of her life caring for her disabled mother. Young enrolled at Howard University in 1916, but it would be another seven years before she would graduate with a bachelor's degree. She then went to the University of Chicago, where she received a master's degree and published her first article in *Science*.[7] She returned to Howard University, where she taught and conducted research with her former professor Ernest Everett Just. For years, Young's name would appear on the grant applications, but not on the resulting publications.[8] Rumors of an affair and confrontations between Just and Young corresponded with her dismissal from Howard in 1936. She used this transition as an opportunity to restart her doctoral work at the University of Pennsylvania. She graduated in 1940, making her the first African American woman to receive a doctorate degree in zoology. After her death in 1964, a group of conservationists established the Roger Arliner Young Marine Conservation Diversity Fellowship to support other African Americans who want to engage in marine environment research.

The Importance of Research Funding

Research funding plays an increasingly prominent role in the success of scientific careers. In previous eras, the receipt of funding was considered an input variable in science: a necessary requirement to conduct research. In recent years, however, the awarding of grant funding is seen as an output variable—that is, as an indicator of success. Researchers receive accolades and reputation from the mere fact of having been awarded a grant, before any research outcomes have been realized. The status of funding is reinforced in evaluation criteria at academic institutions: grant receipt is essential for obtaining tenure at research-intensive universities and greatly increases the probability of obtaining a full professorship.[9] Part of the explanation for this shift is the increasing dependency of universities on external funding, which leads to higher competitiveness of grants. In the United States, a substantial portion of federal funding goes to the university to cover the facilities and administrative costs of running a research institution. Indirect costs, as they are known, became policy effective in 1966 and now average around 55%, a substantial amount of the total award.[10] For example, in 2009, Northwestern University received nearly a third of its revenue in the form of research funding, much of which comes from governmental sources.[11] Grants are, as a consequence, critical for the financial well-being of the contemporary university in the United States. This leads to higher levels of competition for scarce federal resources: the rate of success with US federal agencies is slightly below 20% and has declined over time. It is therefore unsurprising that universities often use the amount of funding received as a marketing tool, demonstrating the research prowess of the institution. Grantsmanship has become an essential skill of the twenty-first-century scientist.

Given the centrality of grantsmanship, it is imperative to understand whether the review and distribution of grants is meritocratic. Evaluation of grant proposals is fundamentally different from other forms of peer review. In journal peer review, for example, reviewers are essentially asked to make binary decisions about the suitability for publication (with or without some degree of revision). In grant proposals, on the other hand, reviewers provide narrative feedback, rate individual proposals on an interval scale, and then often deliberate as a panel in order to construct a ratio or categorical ranking. The scarcity of funding leads to a situation where there are far more qualified proposals than funding available. Reviewers, as a result, are asked to differentiate among proposals of similar "quality" using criteria like "excellence" and "originality," which are characterized by considerable ambiguity and thereby create room for idiosyncratic interpretation.[12]

Concerns about the validity of peer review have emphasized the lack of agreement among grant reviewers.[13] However, many scholars have argued that disagreement is not necessarily a flaw, but possibly a feature of peer review. As sociologist Michele Lamont remarks, drawing from her observations of the deliberations of funding panels and interviews with reviewers, "Evaluation is contextual and relational, and the universe of comparables is constantly shifting. Proposals demand varied standards, because they shine under different lights . . . In panel deliberations, the ideal of a consistent or universalist mode of evaluation is continually confronted with the reality that different proposals require a plurality of assessment strategies."[14] Behavioral scientist Gaëlle Vallée-Tourangeau and colleagues noted that healthy heterogeneity is an artifact not only of the proposals but of the composition of the panelists, who bring the "value of interpretive flexibility and cognitive diversity."[15] Reviewers may also be responsive to government or funder priorities when making decisions, adding another dimension of variability in evaluation.[16] Variation across reviewers and reviews, therefore, is not inherently problematic.

Variation becomes an issue, however, when it systematically disadvantages certain populations—failing to follow the notion of universalism.[17] Several studies suggest that review of grant proposals is plagued with disparity and discrimination, including gender, race, education, and age biases in review and award.[18] Several disparities have been observed for women: women are less likely to apply for grant funding; they are less likely to receive funding when they apply; and they receive smaller grants, on average, than their men counterparts.[19] Furthermore, counter to expectations, it has been found that women's success rate declines with age.[20] Related to (and perhaps as a consequence of) these findings, women are underrepresented in disciplines where research is expensive.[21] As we demonstrated in Chapter 3, men's authorship is disproportionately associated with contributing resources, whereas women's authorship is significantly more likely to be linked with experimentation. One explanatory factor could be the unequal and potentially biased distribution of grant funding.

Medicine is a particularly interesting case, given that women have long been matriculating at similar rates but remain underrepresented in academic positions and in grant receipt. One explanation could be the lower rate of grant submissions: less than a third of National Institutes of Health (NIH) proposals come from women, lower than their rate in the field.[22] Differences in the outcomes of submitted proposals do not emerge at the initial submission but rather in renewals—an essential step for early-career biomedical scientists.[23] A study of these renewals revealed implicit biases in textual critiques about the perceived competence of women investigators.[24] This finding resonates with health informatics researcher Holly Witteman and colleagues' analysis that women fare better when the focus is placed on the content rather than an assessment of the investigator.[25] Taken together, these studies argue for the lingering implicit biases that may be hindering the advancement of women in the biomedical sciences.

There are, however, conflicting results on gender disparities in funding, given that most studies use different data sources and are often focused on single competitions, research programs, or funding agencies.[26] Although comprehensive data sources exist for bibliometric data that allow comparison between countries and most disciplines all at once, no such funding databases exist at the global level. Data are organized by country and are either stored on national research databases or simply not available for research.[27] Furthermore, while data on disparities in attainment can be derived from online information, studying success rates is nearly impossible without direct engagement with the funding agencies. Information on the peer review process itself is also limited to the few who are granted short-term access to limited datasets or periods of observation. Furthermore, despite reporting requirements around issues of gender equity, data on sociodemographic characteristics and grant funding are still widely unavailable.[28] This renders studies of equity difficult and makes it nearly impossible to conduct studies that seek to investigate disparities from an intersectional approach. Intersecting and compound disadvantages arise from many angles. For example, prestige bias—preference for established scholars from well-established institutions—has long plagued grant review.[29] This issue intersects with gender, as women tend to be less likely to have this level of prestige; attrition, therefore, is cyclically reinforcing. Studies in the United States have also demonstrated the significant disadvantage of Black and African American applicants for grant funding.[30] Women of color, therefore, may be particularly disadvantaged in resource allocation. Lacking are comprehensive datasets that integrate information on applicants, awardees, and their subsequent success.

We acknowledge these constraints and address them in this chapter by triangulating data from several sources: bibliometric data, federal funding data, and data from a firm specializing in research evaluation (Academic Analytics). Bibliometric analysis draws on funding acknowledgments found in scholarly articles indexed by the Web of Science (WoS) since 2008.[31] These data, however, are limited in their self-declared nature: funding information indexed by the WoS can only be as precise as what authors supply on published articles.[32] While some authors provide funders' names along with grant numbers, others only provide the name of the funder, without any further information. In no case are funding amounts provided in the metadata, which means that the bibliometric analyses of funding cannot assess the size of grants obtained. Moreover, funding acknowledgment practices vary across disciplines and countries: while funding disclosures are established in most disciplines of medicine and health sciences, this is much less the case in the social sciences and humanities; some countries have strict policies regarding grant acknowledgment (for example, China), while others are much less stringent in their monitoring (for example, Canada).[33] Another limitation is that funding cannot be attributed to specific authors, which means that the units analyzed are articles rather than authors. Despite those limitations, funding acknowledgment data provide a macro-level view of funding at the global level and of the gender differences across disciplines and gender of lead authors. These data provide an original, but limited, view of funding.

To complement the bibliometric data, we draw from a US-centric dataset: Academic Analytics, a firm that provides research analytics to US academic institutions. Despite being at the center of a few controversies regarding the quality of its data,[34] its coverage of funding and scholarly publications remains robust. The database used in the analysis contains 165,241 scholars across 375 institutions in the United States for the 2011–2016 period. Grant data are sourced from thirteen federal agencies and two nonfederal sources; these agencies represent the primary sources of funding for US academics. It is worth noting that grant amounts are annualized: a grant worth $400,000 distributed over four years would be counted as $100,000 in the dataset. Each faculty member is linked to numbers of publications, grants, disciplinary information, and sociodemographic data such as gender and year of doctoral degree. This dataset is particularly useful for analysis because it incorporates funding for the humanities and social sciences, disciplines typically omitted from earlier studies. Furthermore, other sources of public data do not provide disambiguation at the individual level, making it impossible to analyze the average grantsmanship per person.

The lens of Academic Analytics provides strong details on individual-level activities for a brief period. For a diachronic and system-level analysis, we turn to agency-level data for the United States and Canada, examining trends at the National Science Foundation (NSF) as well as for the three main federal funders in Canada: the Canadian Institutes of Health Research, the Natural Sciences and Engineering Research Council, and the Social Sciences and Humanities Research Council and their associated

Canada Research Chairs. These provide two comparative examples of national funding. We acknowledge, however, that the trends observed for these funders may not be reflective of what is observed across all countries or even within a single country (funding practices vary dramatically, for example, between the Department of Defense in the United States and the NSF). Data triangulation begins to mitigate the limitations of individual datasets but begs the development of more comprehensive datasets for future analysis.

A Bibliometric Approach to Studying Research Funding

Of all 27,462,109 articles published and indexed in the WoS since 2008, 44.8% ($n = 12,305,452$) contain funding acknowledgment information. Women are first authors on 35.6% and last authors on 24.7% of articles with disclosed funding, which is lower than their representation in those authorship positions overall (46% and 32%, respectively) (see Chapter 2). Even accounting for their lower rates overall, gender differences remain: 44% of women-led and 49% of men-led articles have funding information (Figure 4.1, left panel). These results suggest that women's research is funded proportionally less often than men's. Rates, however, vary by discipline: women have higher rates of funded papers than men in arts and humanities, as well as clinical medicine, social sciences, engineering, mathematics, and physics. The latter three disciplines are particularly striking, given that women are underrepresented in these domains (see Chapter 1). This may suggest a selection effect at work: in disciplines where women are in the minority, the women who survive are those who outperform men. Men have higher funding in all other disciplines, with the strongest disparities in chemistry, health, and professional fields. As mentioned in the introduction, this has strong implications for the direction of research within these areas.

Some of these differences are explained by variation across fields in the percentage of publications that are funded and gendered differences in authorship rates by field (see Chapter 1). To control for these effects, we normalized the percentage of funded articles of men and women in a given specialty by the overall percentage of funded papers of that specialty (Figure 4.1, right panel).[35] This reduces the observed disparities, but it does not eliminate any of the patterns of gendered advantages across discipline. However, the normalization inverts the overall percentage of

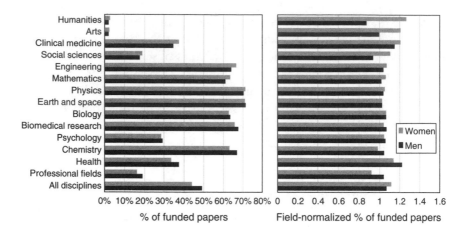

FIGURE 4.1. Percentage of papers with funding disclosure, by discipline and gender of single or last author (left panel); specialty-normalized percentage of papers with funding disclosure, by discipline and gender of last author (right panel), for 2008–2020. Specialty normalization is obtained by dividing each gender's percentage of funded papers in each specialty by the overall percentage of funded papers of that specialty. Normalized scores are then combined into a weighted average by discipline and across all disciplines. When the score is above one, it means that the gender has a higher percentage of funded papers than expected; when it is below one it means the opposite. Specialty normalization includes papers for which the gender of authors is unknown. Those papers are, on average, less likely to contain a funder acknowledgment. This explains why both men and women have, in some cases, a percentage of funded papers above expected values. Web of Science database.

funded papers, demonstrating that, when controlling for the expected percentage of funded papers by specialty, women are, overall, more likely to be funded. This shows that women fare particularly well in under-resourced disciplines, a finding that reinforces earlier studies on the exclusion of women from research with high degrees of resource dependency.[36]

Collaboration is the normative mode of research production, particularly in highly resourced areas (see Chapter 2). Therefore, it is essential not only to examine single- and last-authorship positions but also to understand the composition of teams, by the gender of first and last authors. Using normalized funding rates, we find that teams with men in senior positions (last authorship) and women in lead roles as early-career researchers (first author) are the most likely to have funded papers; the least likely are those with men in both lead positions (Table 4.1). It is notable that those with women in both lead authorship positions are funded at nearly identical rates to those with men in senior and women in junior

TABLE 4.1. Normalized percentage of papers with a funding disclosure, by NSF discipline and gender composition of author leadership positions, 2008–2020. Normalization is obtained by dividing each gender's percentage of funded papers in each discipline by the overall percentage of funded papers of that discipline, and aggregation at the level of all disciplines is weighted by the percentage of papers in each discipline. When the score is above one (darker cells), it means that the gender has a higher percentage of funded papers than expected; when it is below one (lighter cells) it means the opposite. M-M: men first and last authors; M-W: man first author and woman last author; W-W: women first and last authors; W-M: woman first author and man last author. Web of Science database.

Discipline	M-M	M-W	W-W	W-M
Humanities	0.89	0.96	1.18	1.06
Arts	1.00	0.83	1.01	1.11
Professional fields	1.02	0.94	0.98	1.04
Social sciences	0.95	0.97	1.04	1.11
Psychology	0.99	0.91	1.01	1.05
Clinical medicine	0.92	1.01	1.12	1.08
Health	1.00	0.96	0.97	1.07
Biology	0.99	0.99	1.00	1.03
Engineering	0.98	1.04	1.05	1.04
Mathematics	0.99	1.01	1.03	1.02
Chemistry	1.01	0.98	0.97	1.01
Biomedical research	1.00	0.99	0.99	1.02
Physics	0.99	1.02	1.03	1.02
Earth and space	0.99	0.99	1.00	1.03
All disciplines	0.97	1.00	1.05	1.05

positions. Furthermore, in clinical medicine, physics, and engineering, articles with women in both dominant positions are more frequently funded, on average, when controlling for expected values by specialty. The interpretation of those results is not straightforward. While they could mean that women have higher rates of funding, the fact that this analysis is done at the article level—examining the proportion of articles with funding information—could also suggest that women are more efficient with individual grants (such as producing more articles on average, with a smaller amount or number of grants).

The probability that an article is funded varies as a function of team size (Figure 4.2), with the likelihood of funding increasing with team size and stabilizing around seven coauthors. As shown in the left panel of Figure 4.2, collaborations with men as both first and last authors exhibit the highest level of funding for smaller teams; yet their percentage of funded papers is the lowest for teams that have more than eight authors. On the other hand, teams led by women as first and last authors exhibit the lowest percentage of funded papers for smaller teams, and their percentage of funded papers is the highest within the largest teams.

This difference is even more striking when the discipline of articles is considered through normalization (Figure 4.2, right panel): articles with women as both first and last authors consistently exhibit higher levels of funding, while articles with men in those two dominant positions show

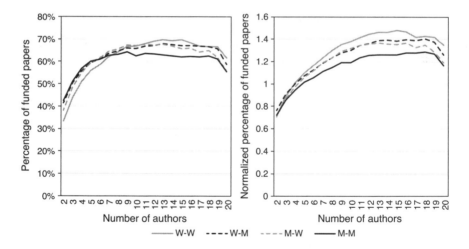

FIGURE 4.2. Percentage of papers with funding disclosure, by team size and gender composition of author leadership positions (left panel); normalized percentage of papers with funding disclosure, by discipline and gender of last author (right panel), for 2008–2020. Normalization is obtained by dividing each gender's percentage of funded papers in each specialty by the overall percentage of funded papers of that specialty. Normalized scores are then weighted and aggregated at the level of discipline and of all disciplines. When the score is above one, it means that the gender has a higher percentage of funded papers than expected; when it is below one it means the opposite. Only papers with more than one author are considered. Web of Science database.

the lowest levels of funding. Papers with mixed leadership teams are in between. These findings imply that papers led by women are more likely to acknowledge external funding. Combined with the results presented earlier on gender differences in research productivity—which showed that men were more productive than women (Chapter 1)—these findings reinforce the argument of efficiency: women are more efficient at translating grants into research articles. This is also supported by research on the reinforcing relationship between research funding and outputs, which shows that a large proportion of the research productivity puzzle is explained by disparities in research funding and that, at equal funding rates, women are as productive as men.[37]

Gender Differences in US Funding

Bibliometric databases only illustrate, in a binary way, whether articles led by men and women are funded. They cannot detail whether men or women are more likely to receive funding when they apply and how much funding they receive. For this information, we need to move to country-level datasets that compile funding success rates and funding amounts. We begin with an analysis of the United States, using the Academic Analytics database. This analysis provides an overview, at the level of individual researchers, of funding across the primary agencies.[38] While a direct comparison with the worldwide data compiled from the WoS is not possible—given that we are comparing global papers to country-specific funding—some parallels can still be drawn.

Figure 4.3 (left panel) presents the percentage of researchers from US universities who have received at least one grant, by NSF discipline.[39] While the results are not perfectly aligned with those obtained for WoS funding acknowledgment, they converge on one important aspect: women are more likely to be funded in fields where they account for a small percentage of researchers, such as engineering, physics, earth and space sciences, and, to a lesser extent, mathematics. All disciplines combined, however, men are more likely to obtain at least one grant over the period (41% versus 35%).

A similar trend is also observed for the total grant award received (Figure 4.3, right panel): while the annual average total grant size received by men is $220,000, it is around $170,000 for women. Such differences are more pronounced in the biomedical sciences: clinical medicine, biomedical

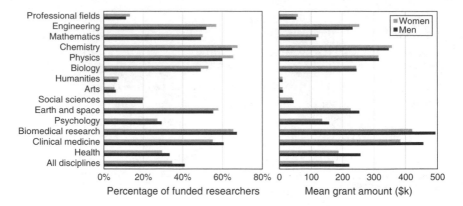

FIGURE 4.3. Percentage of researchers with at least one grant, by NSF discipline and gender (left panel); mean total amount of grant received, by NSF discipline and gender (right panel), 2012–2016. Academic Analytics Database. Figures A.3 and A.4 provide distribution of women researchers by decile of funding and discipline; those confirm the results obtained by the averages.

research, and health are the three disciplines with the highest gendered differences in total funding received. These findings align with the publication data, suggesting that women are less likely to receive funding, receive less funding on average, and publish fewer funded articles as a result. This is cyclically reinforcing, as having fewer publications will likely lead to a lower probability of future grant success.

A different portrait is obtained in several disciplines of the natural sciences: women receive slightly higher grant funding than men in engineering, mathematics, and chemistry, and they are on par with men in physics. The finding in engineering resonates with previous research and aligns with the bibliometric data, which demonstrates that engineering and fields with the highest proportion of funded articles are also those where women are disproportionately likely to have funded articles.[40] This reinforces the argument of the selection effect: only funded women remain in these disciplines.

As with publications, age plays a significant role in grant funding.[41] Figure 4.4 presents the funding information of US scholars covered in the Academic Analytics data as a function of the year in which they obtained their last degree. The data show that older faculty (PhD before 1975) are much less funded than scholars who obtained their PhD twenty to thirty years ago, and the mean funding rates plummet for those who obtained

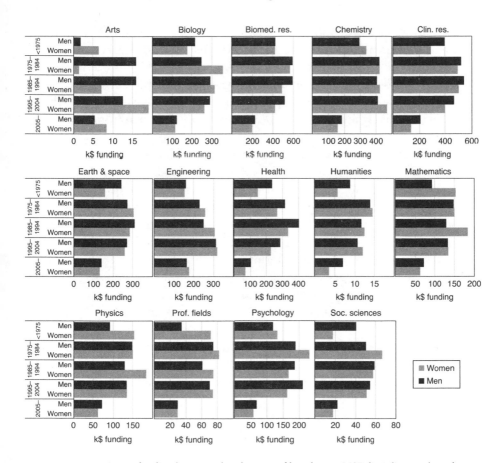

FIGURE 4.4. Mean funding by researcher, by year of last degree, NSF discipline, and gender, 2012–2016. Academic Analytics Database.

their PhD from 2005 onward. This can have problematic implications in multiple ways: it demonstrates that funding is more likely to be received by those with secure positions in academia, thus reinforcing the Matthew effect (see Chapter 2). In a context where the age at first award has increased, this is problematic for junior scholars, particularly when grant funding is increasingly used as an impact variable for consideration in tenure and other awards.[42]

As shown in Figure 4.1, gender differences in funding for chemistry, engineering, mathematics, and physics are minimal, and both genders obtain similar rates of funding across career ages. However, for biomedical research, clinical medicine, and health, a disparity in funding received by men and women is visible at each career stage. In clinical medicine, this difference increases as a function of career age, with the largest gap observed for the most recent graduates. A few factors may explain these trends, such as survival bias: older women who are still active have had to work in academic climates that were much less favorable to women—this makes them likely to have as much symbolic capital as men from the same cohorts.

While the extensive coverage of Academic Analytics allows us to analyze gender differences across a broad array of funding sources, it lacks historical depth—focusing primarily on contemporary data. Data from the NSF allow us to compare, over time, the proportion of funding received by men and women, both in terms of standard and continuing grants—which account for 68.6% of all funding awarded by the NSF over the 1980–2020 period—and in terms of fellowships, which are primarily given to junior scholars (Figure 4.5). These data show that, until the end of the 1990s, women's grants were much smaller than those of men, with women obtaining percentages of men's grants that oscillated between 62% and 94%.

However, from 2001 onward, women obtained mean grant amounts that were slightly above those of men. These differences are mostly due to the NSF directorates in which women are more dominant (for example, Education and Human Resources). In terms of proportion of total funding, we observe that, although men still account for a larger share of the funding—which is not surprising given their higher proportion of the workforce—their share is significantly decreasing, from 95% in 1981 to less than 70% in 2020. In terms of fellowships awarded—which includes both graduate and postdoctoral—we also observe that, while men still account for most funded projects, women are nearly at parity. More specifically, while women's share of fellowships was less than 10% in the early 1980s, they accounted for more than 47% in 2020.

The relative percentage of funding obtained by women varies greatly across the seven directorates of the NSF (Figure 4.6). While an upward trend is observed across all directorates, the relative degree of funding obtained by women in each of them varies greatly. The Directorates for Education and Human Resources and for Social, Behavioral, and Economic Sciences have the highest percentages of their funding allocated to women, increasing from about 20% in 1980 to more than 50% in 2019 (Education and Human Resources), and from slightly above 10% in 1980 to 45%–50% in recent years (Social, Behavioral, and Economic Sciences).

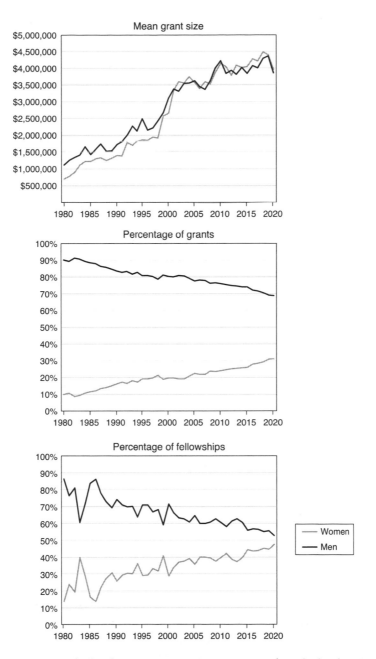

FIGURE 4.5. Mean standard and continuing grant size, percentage of standard and continuing grants, and percentage of fellowships at the NSF, 1980–2020.

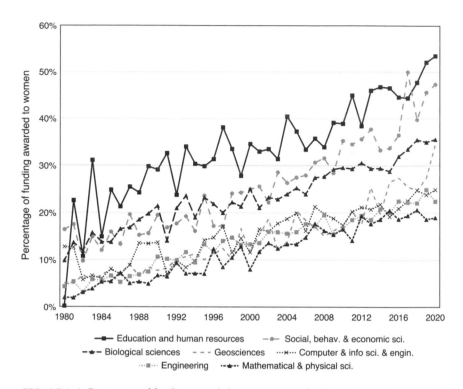

FIGURE 4.6. Percentage of funding awarded to women, standard and continuing grants, by directorate, NSF, 1980–2020.

The trend observed for Biological Sciences was similar to that of Social, Behavioral, and Economic Sciences until the early 2000s, when the latter increased at a slower pace. All four other directorates follow a similar pattern, increasing from about 5% in 1980 to 20%–25% in 2020—except in Geosciences, where the percentage of funding obtained by women increased to 34% in 2020.

Grant size remains higher for men than for women, despite improving after 2000 (Figure 4.7). Mean amounts obtained by men were more than 30% higher than those obtained by women in Computer and Information Science and Engineering, Mathematical and Physical Sciences, and Social, Behavioral, and Economic Sciences between 1980 and 2000; this gap reduced significantly after 2000 and oscillates between 5% and 11% for those three directorates. A sizable gap was also observed in Biological Sciences (10%) and Geosciences (19%); it reduced to 3% and 10%, respectively. A gap lower than 10% was observed in Education and Human

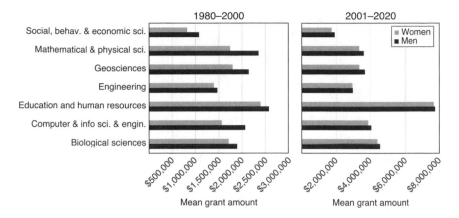

FIGURE 4.7. Mean grant amount obtained by women and men, standard and continuing grants, by directorate, NSF, 1980–2020.

Resources and in Engineering; it decreased to less than 1% after 2000. Interestingly, the two domains with no gap in research funding are the two with the highest extremes in terms of their percentage of women in the workforce: Engineering, which is among the lowest, and Education, which is among the highest (see Chapter 1). For Engineering, the absence of a gap in mean amounts received may be due to survivor bias, by which the women who remain in this heavily men-dominated sector must over-perform.[43] For Education, this suggests that, when parity is reached in terms of women's participation, resources become more evenly distributed across genders. Taken combined, these results suggest a slow march toward parity but also indicate that interventions may be needed, particularly in STEM fields, to reach parity.

Canadian Funding across Disciplines

In Canada, research funding is awarded through three major research councils: the Canadian Institutes of Health Research (CIHR), the Natural Sciences and Engineering Research Council (NSERC), and the Social Sciences and Humanities Research Council (SSHRC). Combined, these councils have a budget of slightly more than Can$4 billion annually and are responsible for the majority of external research funding in the country. They provide funding to faculty to cover the direct costs of research (salary and benefits for research personnel, stipends for trainees, travel, and equipment), as well as provide direct funding to graduate students in the form of scholarships, which are tax exempt. Although several provinces (Québec, Alberta, and British Columbia) also have provincial councils, their budgets are marginal compared with those from the federal government. Funding databases from the federal councils provide an opportunity to analyze historical trends in women's share of graduate scholarships and grant amounts, standard research grants for research, and Canada Research Chairs—a prestigious program that funds about 2,000 research professorships in all disciplines across the country.

We observe that, in the social sciences and humanities (SSHRC) and medical sciences (CIHR), women obtain a higher percentage of all graduate and postdoctoral scholarships (Figure 4.8). However, in natural sciences (NSERC), their percentage remains under 40% throughout the period, and it has declined between 2004 and 2014, after which it started to increase again. The decline—followed by a rebound—may be due to the ten-year reign of the Conservative Party, which was considered to be less sensitive to issues of gender equity than the Liberal Party, which was elected in 2015.[44] Overall, the results suggest a relatively strong investment in junior women scholars in Canada.

The same level of funding is not observed for resources garnered later in academic life—the percentage of grants obtained is lower than for scholarships. This is both an indicator of and a contributor to attrition in the scientific workforce. That is, there are fewer women in higher ranks, which explains the lower proportion of funding to women, and the lack of resources for women in midcareer stages may also lead to greater attrition. There are, however, gains in this arena over the last two decades: the percentage of funding obtained by women increased by 123% for NSERC, 70% for CIHR, and 31% for SSHRC. Since 2017, women and men have been at parity in SSHRC funding.

The Canada Research Chairs program provides exceptional opportunities for scholars to accelerate their research agenda. Although the percentage of Canada Research Chairs for women has increased, it remains much lower than what is observed for both scholarships and grants, which emphasizes the higher gender inequities at the level of prestigious awards. The figure also shows a decline in the proportion of funding for women at the end of the first decade of the 2000s, in a manner similar to what is observed for scholarships. The sharp rise observed after 2016 can be attributed to aggressive policies and quotas imposed by the federal

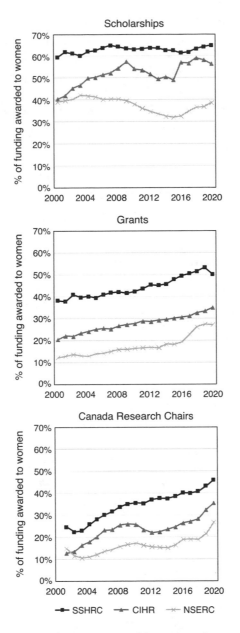

FIGURE 4.8. Percentage of amount received by women, by type of grant (scholarships, grants, Canada Research Chairs) and research council, 2000–2020.

government on the program.[45] These variations according to political interventions demonstrate that governments can make actionable policy toward gender equity in science.

Despite an increase in the percentage of overall funds received by women, there are strong differences in the size of grants received for medical science (CIHR) and natural sciences and engineering (NSERC) (Figure 4.9). Men obtain consistently higher grant amounts at CIHR, with

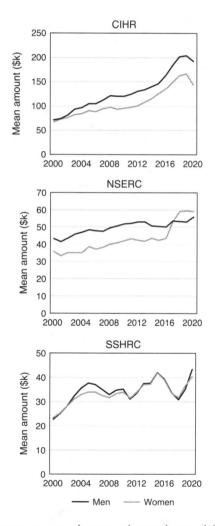

FIGURE 4.9. Mean amount per research grant and research council, by gender, 2000–2020.

the gap increasing from 6% in 2000 to 34% in 2020. An opposite trend is observed for NSERC: while men's grants were more than 20% higher than those of women in 2000, women's grants reached the level of men's in 2017 and have been slightly higher than men's since then. These changes may be related to policy interventions: in 2015, NSERC committed to implementing the Status of Women Canada Departmental Action Plan on Gender-Based Analysis.[46] In the case of SSHRC (social sciences and humanities), the mean amounts received by men and women remain the same throughout the period, although men obtained slightly higher grant amounts in the first decades of the 2000s.

The success rates of men and women at obtaining funding are crucial elements to consider when analyzing gendered differences in funding.[47] While this information has historically not been available, Canadian research councils have recently provided information for most of their programs. CIHR provides success rates and various other funding-related indicators by gender of the principal investigator.[48] Despite obtaining smaller grants, women's success rates are similar to men's. This reinforces previous research in biomedicine: women's lower funding overall is associated to smaller funding requests—in terms of both amounts requested and number of proposals—rather than to lower funding rates.[49] These trends vary, however, by competition year and age group (Figure 4.10). While in the early 2000s women had slightly lower success rates, the trends reversed at the end of the first decade, with women researchers having a slightly higher success rate. This appears related to an overall decline in grant applications' success rates: men had comparatively greater chances of being funded when success rates were higher, while women had relatively higher success rates when rates were lower.

The data also show interesting age-related patterns. While there are few differences in men's and women's success rates for younger cohorts (forty-five years old or less), there is a significant difference for older age groups, with much higher probability of being funded for older men. Another striking feature is the relatively stable success rate for women throughout age groups, which remains at about 11%–12% until age sixty-five, while success rates increase for men as they become older. Analysis of the ratio between funds approved and funds requested shows that, throughout competition years and research areas, women are more likely to obtain a higher percentage of the amount requested than men.[50] While this is undoubtedly related to the fact that women ask for significantly less funding per grant than their men colleagues, it also

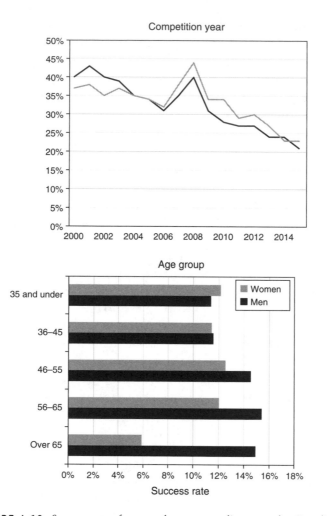

FIGURE 4.10. Success rate of men and women applicants at the Canadian Institutes of Health Research grant competitions for 2000–2001 to 2015–2016 competitions by competition year (top panel) and by age group (bottom panel) for the 2015–2016 competition. "Gender Equity Data Analysis—CIHR Competition Success Rates by Gender (All CIHR Grant Programs)." Canadian Institutes of Health Research, accessed July 28, 2022, http://www.cihr-irsc.gc.ca/e /50070.html; "Gender Equity Data Analysis—CIHR Competition Success Rates by Gender (Project Grant)," Canadian Institutes of Health Research, accessed December 24, 2021, http://www.cihr-irsc.gc.ca/e/50235.html.

could imply that women's expected budgets are closer to the actual cost of research.[51]

Strategies for Equality in Science Funding

Funding is essential for success in the contemporary scientific environment. Yet we find gendered disparities in funding, with women less likely to have received a grant, representing fewer authorships among funded papers, and having a lower proportion of their papers with funding information. This leads to cycles of disparity: when women's work is not funded, they produce less, thereby leading to lower rates of funding.

Observed disparities, however, are complicated once one accounts for differences across and within disciplines. Our results suggest that, in disciplines where women are underrepresented, the gender gap in funding is reduced, and that in many cases, women have higher rates of funded articles, higher rates of funding, and higher mean grant rates.[52] We also find that women outperform expected values, producing higher levels of funded articles than would be expected given the general rates of funding across many specialties. This is reinforced by the Canadian data, which show that men had a comparatively greater chance of being funded when success rates were higher, while women have relatively higher success rates in times of scarcity. It may also be a selection effect, particularly for fields hostile to women, where only those with high rates of funding remain. These results suggest that policies should examine the idiosyncratic nature of disciplines: to address early attrition effects in highly masculinized fields and to focus on funding amounts, particularly for midcareer scholars, in fields with higher degrees of parity.

Several studies have reported a relationship between age and funding application and success. In the United Kingdom, women applied for as many grants as men before age forty and were more successful; however, those over fifty years old applied for fewer grants and were less successful in their applications.[53] Studies of Québécois scholars have shown a decline in the amount of funding women receive after age thirty-eight.[54] In the United States, the average age for someone receiving an NIH R01—a critical requirement for tenure in biomedical disciplines—is forty-two, which is within the age category that women are more likely to be at a disadvantage.[55] Moreover, the precursor to receiving this grant is the postdoctoral training grants given by the NIH, and women experience a 20%

lower rate of transition from these postdoctoral grants to the R01 grants.[56] While differences in productivity explained some of this gap, women with similar publication records still transitioned to the lead grants at lower rates than men. However, after the first grant, women have similar funding longevity to men.[57] This reinforces that, in the United States, a critical moment is the initial R01 funding moment. The lack of women in biomedicine has strong implications for women's health research, as discussed earlier.[58] This makes the widening gap at CIHR and the results from the NIH particularly sobering.

One may argue that these results will improve over time, as more women progress in their careers and take on positions as principal investigators. The data, however, do not support this. If we look at the proportion of men and women who have received a grant, as a function of the decade in which they received their degree, we see striking disparities in the youngest cohorts and virtually no difference in the oldest cohort. In other words, the funding gap is larger with younger cohorts than it is with older cohorts, which suggests that the current growth in the percentage of women in the system will not solve this issue. This may imply a survivor bias, in which the women who remain in academia are those who were able to receive grants at rates similar to those of men in their cohort. The gap in junior scholars may be an early indication of later attrition, given the increasing pressures on faculty to receive at least one large grant before tenure.[59]

Funding agencies have a critical role to play in guaranteeing that principal investigators are representing the wide range of identities and needs in the population—ensuring equity not only by gender but along all axes of underrepresentation.[60] In order to move toward equality, we must analyze the mechanisms underlying the distribution of resources. Peer review is a critical mechanism underlying some of these disparities. As noted earlier, it has been demonstrated that women suffer in the grant review process, particularly when the scientist is reviewed rather than the science.[61] An analysis in the Netherlands showed that peer review systematically disfavored women applicants, and that equity in the outcomes could only be achieved through agency intervention.[62] Changing the demographics of review panels might be a simple solution to this problem: one study demonstrated a relationship between the diversity of those awarding grants and the awardees.[63] Similar results were also obtained at the level of journal submissions.[64]

Changes in the composition of gatekeepers are happening, albeit slowly. In the United States, astronomer France Córdova served as the fourteenth director of the NSF, following two other women (microbiologist Rita Colwell [1998–2004] and sociologist Cora B. Marrett [2010; 2013–2014]). In Canada, biochemist Mona Nemer was named in 2017 as the country's chief science advisor, and all three scientific directors of the Fonds de recherche du Québec are currently women.[65] In South America, Argentinian research council CONICET is presided over by Ana María Franchi, a biological chemist, and legal scholar Aisén Etcheverry heads ANID, the Chilean national agency for research and development. Microbiologist Sue Thomas has served as the CEO of the Australian Research Council since 2017. Sociologist of science Helga Nowotny served as the second president of the European Research Council. Fiona Watt is the CEO for the Medical Research Council in the United Kingdom: her appointment in 2018 made her the first woman CEO of this body, which was established in 1913. Some councils are still lagging. The National Environment Research Council in the United Kingdom was formed in 1965 and has yet to see a woman at the helm.

Funding agencies are in a prime position to increase equity, both in directly increasing funding for women researchers and in incentivizing cultural shifts toward equity. Two funding initiatives have been particularly successful in addressing gender disparities in funding and motivating institutional responses in policies and practices: Athena SWAN (Scientific Women's Academic Network) in the United Kingdom and ADVANCE in the United States. The Athena SWAN charter is owned by the Equality Challenge Unit in London.[66] The unit recognizes institutions that provide evidence of practices that address gender inequalities. Based on submitted documents, institutions are characterized at three levels: bronze, silver, or gold. In 2011, the National Institute for Health and Care Research in the United Kingdom announced that it would require applicants to have at least silver status. The result was a tenfold increase in the number of silver or gold awards in 2016, compared with 2011.[67] Furthermore, there was a 24% increase in women leads in 2016, from 8% in 2011. Notably, only one of the awards was gold, demonstrating that institutions were working toward the stated benchmark (and not above it). This may suggest a sort of incentive-matching (but not exceeding) gamesmanship, but one that the funding agencies can leverage toward stated goals of equity.

Institutional transformation is the focus of the ADVANCE program in the United States, which began in 2001 at the NSF as a response to a 1997 workshop that analyzed the NSF's portfolio and recommended "structural approaches to promote systemic change and gender equity."[68] From inception to 2018, the NSF awarded nearly $300 million in Institutional Transformation grants to more than 179 higher education and non-profit institutions in the United States. Analyses have shown that those institutions in the initial cohorts of grantees increased in the proportion of women faculty at greater rates than comparable institutions not awarded these grants.[69] ADVANCE has also begun to follow the practices of Athena SWAN by providing institutional ratings. The STEMM Equity Achievement (SEA) Change initiative (led by the American Association for the Advancement of Science) catalyzed in 2016 with the goal of providing bronze, silver, and gold ratings to institutions for best practices in pursuing equity.[70] The first set of bronze awards were given in 2019; at present, five institutions have achieved this level.[71]

Individual funding programs also have the power to initiate change. In 2017, the Canada Research Chairs program implemented the Equity, Diversity and Inclusion Action Plan, which set targets for institutions in terms of the composition of their awardees, to be met within eighteen to twenty-four months.[72] Each institution was required to have a certain percentage of its chairholders meet their criteria in terms of a set number of women, people with disabilities, Indigenous peoples, and members of visible minorities, and those targets were determined by the disciplinary composition of its university community, as well as of the population of the region where it is located. Institutions that did not meet the requirements of the action plan experienced penalties, such as the withholding of peer review decisions until requirements had been met.[73] While the vast majority of institutions satisfy the criteria of the action plan, a certain number of smaller institutions have not yet done so, which indicates that additional resources may be needed to implement these policies.

Many scholars have recommended that funders adopt an "institutional report card for gender equality" like Athena SWAN and SEA Change.[74] However, the burden for reporting often falls on women faculty. In fact, in the wake of the pandemic and the increased burdens it placed on women, the National Institute for Health and Care Research removed the Athena SWAN requirement for awards in 2020. Bibliometric data may provide one solution for this dilemma. Indicators such as the rate of women authorship could be used to provide benchmarks to meet funding mandates, in lieu of more burdensome processes. We have argued that these types of "indicators for social good" can function as a new mechanism in responsible metrics.[75] There has been some uptake by rankers: the proportion of women authorship is now provided as a variable by which one can sort institutions in the Leiden Ranking.[76]

These types of initiatives and interventions are necessary to accelerate parity in funding and move toward equity in science. As noted earlier, funding is increasingly important in all realms of academic life: receiving funding is both an explicit and implicit requirement for career advancement. It is a prerequisite for research activity, particularly in fields where research is highly collaborative and internationalized. Participation in a collaborative and global scientific environment costs money. Studies have shown that about than 30% of all articles have a graduate student as a first author, and more than 40% have a postdoctoral researcher as the lead author.[77] As we have shown, articles with men in senior and women in junior roles are the most likely to be funded, reinforcing power dynamics. It is essential that funding be directed toward junior scholars directly and toward senior women in mentorship positions. Same-gender mentoring for women has synergistic benefits, both to students (as mentioned in a previous chapter) and to advisers, who gains access to strong collaborators. These, however, come at a considerable cost, particularly in countries where funding is used to provide tuition, stipends, and health insurance.[78] Funding is also necessary to provide opportunities for these students and for the faculty members to travel to conferences, research sites, and other institutions in order to extend collaborations and the impact of their work. Mobility is a luxury of the wealthy, even in science. We move, therefore, to examine gender differences in mobility.

Chapter 5

Mobility

In the late nineteenth century, German doctoral degrees had a high reputation around the world, and many American men traveled abroad to receive these prestigious credentials. Christine Ladd-Franklin (1847–1930) believed that if American women were able to obtain doctorates in Europe, it would create pathways for them back home. Born in Connecticut, Ladd-Franklin had a strong educational pedigree: she was valedictorian at the coeducational Welshing Academy, went to Vassar College to study with astronomer Maria Mitchell, and then moved to Johns Hopkins, where she was given a special dispensation to attend lectures in mathematics (although she was not added to the university's register, as women were not allowed admittance).[1] After four years of study and a published thesis, she was not awarded a degree. Ladd-Franklin turned her attention to German universities, where foreign women were allowed to study, despite a policy against admitting German women. As Alice Hamilton, who studied in Germany in the mid-1890s, recalled, "We were told that the only reason women wanted a university education was to make trouble for the government. If foreign governments did not object, that was all right, but Germany had more sense."[2] The situation in Germany provided opportunities to study, but limited opportunities for degrees. For that, women went to Switzerland.[3] Mobility to and within Europe was necessary for women of the era to receive both the education and credentials they needed.

Ladd-Franklin worked with another academic woman, Ellen Richards, to create the European Fellowship from the Association of Collegiate

Alumnae, which provided funds for women to study abroad. The initial proposal, made in 1888, was for $500 per year for study at a European university. The fellowship was approved in 1889 and the first fellowships were award in 1890, after a year of fund-raising.[4] Ladd-Franklin did not benefit from the fellowships she established but traveled with her husband and daughter to Germany to study with leading scholars. She spent two years in study, culminating in the development of a new theory of color vision, which she presented at the International Congress of Experimental Psychology in London in 1892. Despite her accolades, academic appointments eluded Ladd-Franklin. Her marriage made her ineligible for positions at women's colleges, and she never received more than annually renewed appointments at Johns Hopkins and Columbia. This partially may have been the result of a lack of a formal degree. It was not until she was seventy-eight years old that she was invited back to Baltimore to receive her doctorate from Johns Hopkins—forty years after she finished her thesis.[5]

A few decades later, the mobility pattern across the Atlantic reversed. For example, Ada Isabel Maddison (1869–1950) was a British mathematician whose academic career was primarily spent at Bryn Mawr in the United States. Maddison attended lectures at Cambridge and passed the Cambridge Mathematical Tripos Exam, earning a first-class degree, but it was not awarded, since women were unable to receive degrees from Cambridge. Instead, she was awarded a bachelor's degree from the University of London in 1893. In the final years of her studies, Maddison received scholarships to study at Bryn Mawr and then the University of Göttingen in Germany. She received a doctoral degree from Bryn Mawr in 1896 and continued as a Reader in mathematics. By 1904, she had been appointed as an associate professor and assistant to the president, and she remained at Bryn Mawr until her retirement in 1926.

During her stay in Germany, Maddison met with fellow student Grace Chisholm Young (1868–1944), also an English mathematician who, like Maddison, obtained a first-class degree on the Mathematical Tripos Exam. Chisholm Young additionally obtained a first-class degree at Oxford and took the second part of the Mathematical Tripos at Cambridge. Despite these accomplishments, it was impossible for her to receive a degree (like Cambridge, Oxford did not award doctoral degrees to women at that time). Therefore, Chisholm Young went to the University of Göttingen, where she became one of the first women to be awarded a doctorate in Germany in 1895. Physicist Margaret Maltby also graduated from Göttingen in 1895, supported by the fellowship established by Ladd-Franklin.

When Germany began awarding doctoral degrees to women, it was only for foreign women. As historian Margaret Rossiter notes, "It was German women the German professors were most adamantly opposed to admitting in the 1890s. Foreign women were far less of a threat, since they would return home and not expect to teach in Germany. Nevertheless, the acceptance of foreign women helped set a precedent upon which German feminists capitalized later."[6] Chisholm Young paved the way for future generations of women across the world. In 1902, German women were admitted and began receiving graduate degrees. Despite these advances, Chisholm Young was still a product of her times. She married William Henry Young, a mathematician who had served as her tutor at Cambridge. She wrote hundreds of papers in collaboration with her husband, published jointly or under his name alone. It took nearly twenty years after her doctoral degree for Chisholm Young to publish under her own name and begin to establish an independent scientific identity. Mobility, as we will show, is deeply linked with productivity, yet these stories (and the data provided in this chapter) demonstrate the highly interconnected web of social and scientific factors that influence scientific careers.

Mobility and Scientific Careers

Mobility is deeply woven into the fabric of science. In the United States, for example, one-third of faculty in scientific and technological fields are foreign born and educated, and a sizable proportion of US Nobel laureates are foreign born.[7] Historians of science have focused on the role that war plays as a catalyst in scientific mobility, as seen, for example, in the large influx of scientists to the United States in the throes of World War II. Italian-born neurobiologist Rita Levi-Montalcini, for example, left Italy in 1946 for Washington University for a one-semester research fellowship. She stayed for twenty years, though she was able to establish a second lab in Rome in the 1960s. Czech mathematician Olga Tassky-Todd left Vienna for the United Kingdom in 1935 and then moved to the United States in 1945. Mobility, however, did not provide a safe haven for all women scientists in time of war. Elise Richter (1865–1943) was a Viennese philologist and the only woman to hold an appointment at an Austrian university before World War I. She was one of the first women (in 1901) to be awarded a doctorate from the University of Vienna, the first *Dozentin* (lecturer) at that institution (1907), and the first extraordinary professor (1921). In 1920 Richter became the chair of the Association of Austrian Academic Women, and in 1922 she founded the Austrian Federation of University Women. During World War II, she was deported to Theresienstadt concentration camp in German-occupied Czechoslovakia. She arrived on October 9, 1942, and died on June 21, 1943.

World War II is not the only conflict that led to the loss or displacement of scientists. In the wake of the Maoist revolution of the late 1940s, Flossie Wong-Staal and her family fled to Hong Kong. At the age of eighteen, she moved to the United States and established her career in virology. Ethiopian-born chemist Sossian M. Haile's family fled Ethiopia during the coup in the mid-1970s. They moved to the United States, where Haile did all her degrees, though she took advantage of mobility programs like Humboldt and Fulbright to work in Germany for short periods, demonstrating that mobility is not always a push but also a pull factor.

Research itself can also propel mobility. Science took Merieme Chadid from her hometown of Casablanca, Morocco, to study in France at the University of Nice and then to Paul Sabatier University in Toulouse, where she received her doctoral degree in 1996. Her journey continued to Chile, where she worked on the installation of the Very Large Telescope. In 2005, she took a bigger step: becoming the first Moroccan and first woman French astronomer to travel to Antarctica, where she worked on the installation of a large astronomical observatory. Antarctica also called other women to leave their homes for science. In 1983, geologist Sudipta Sengupta and biologist Aditi Pant were the first two Indian women to visit Antarctica. Sudipta obtained a doctoral degree in India before taking a postdoctoral position at Imperial College in London. In 1977, she joined the Institute of Geology of Uppsala University in Sweden, and she returned to India in 1983. Her research has also taken her to the Scottish Highlands and the Scandinavian Caledonides. For US-born Barbara McClintock (1902–1992), it was South American maize races that gave her the genetic information she needed for her work in cytogenetics. Her work was recognized with a Nobel Prize in Physiology or Medicine, making her the only woman to receive an unshared Nobel Prize in this category.

As exemplified by these anecdotes, mobility can support scientific careers in many ways: providing access to resources, collaborators, or research objects that lead to opportunities for individual researchers and innovations in science.[8] Given the benefits observed, many scientific institutions have begun to incentivize and celebrate mobility to the extent that it has become a new indicator of excellence and an essential component for career advancement.[9] These investments, however, are asymmetrical across the globe and with different obligations for "return mobility." In the case of China, for example, young researchers are sent abroad for education and often expected to come back to their home country after their training.[10] The expectation of mobility is particularly true for early-career researchers; mobility has been deemed "an indispensable element in the career trajectories of young academics"[11] and a "rite of passage."[12]

One could argue that all researchers are mobile, but it is all a matter of degree and distance. Throughout their studies and careers, researchers often move from one institution to another; this can be another institution in the same city, within the same country, or in a different country. Other types of mobility are temporary events: researchers often travel abroad for conferences and research, and established researchers might spend extended time at other institutions as invited professors. Of course, these mobility events do not have the same consequences, but they are nonetheless markers of transitions in researchers' careers. When mobility events move human resources across countries, they also serve to restructure the global science and engineering system. As economist Richard Freeman argues, there are five key components in the globalization of knowledge, almost all of which are intrinsically tied to mobility: the growth of higher education worldwide, the increasing numbers of international students, immigration, nonimmigration trips, and collaboration.[13] The essential nature of academic mobility has consistently been reinforced in academic discourse, despite advances in telecommunications.[14] However, promotion of international mobility at any career stage may come at a cost to the advancement of gender equity.

Qualitative studies have shown that women are less mobile and more likely to sacrifice their careers for the mobility of their partners.[15] There are, however, several nuances to this. For example, younger women tend to be just as mobile internationally as men, and sometimes more so, but this matched level of mobility generally diminishes as they age.[16] There are also differences in types of mobility, between long-term migration and circular patterns of mobility (such as temporary mobility).[17] Women engage in circular mobility at rates similar to men, but their mobility is less frequent, for shorter durations, and closer to home.[18] They are also more constrained by "tied mobility"—that is, to have a domestic partner whose movement may reduce their ability to move. These gendered patterns of mobility are essential for considering attrition in science, particularly when mobility is expected or incentivized.[19] However, we lack comprehensive and contemporary data on scientific mobility.

Data on the mobility of the highly skilled workforce are compiled by national statistical agencies and by international organizations (such as the Organisation for Economic Cooperation and Development and the United Nations Education, Scientific and Cultural Organization), yet they suffer from several compatibility issues: mobility is monitored in different manners in each country or by each agency, and therefore results are often not comparable, either across countries or over time.[20] Analyses have also largely focused on the mobility of students, missing mobility events that may happen at the postdoctoral level or later in the career, such as once researchers have obtained long-term positions.[21] Moreover, data for these studies are generally obtained through surveys, which are not performed on an annual basis, rarely allowing for longitudinal analyses, and are quickly dated.

This chapter draws on a relatively new source to measure researchers' mobility: indexed scientific publications and their metadata.[22] Citation indexes such as the Web of Science increasingly include information on the affiliations of authors, as provided by the authors. Affiliation data can have various levels of granularity—including, for example, information on the lab, center, department, or school. Institution information is provided in nearly every case, though the idiosyncrasies in reporting require significant cleaning for comprehensive studies of mobility. Country information, on the other hand, is a relatively straightforward data point provided in affiliations that allows for rich analysis of global scientific mobility.

In order to compile these data, one must not only identify a country of affiliation but be able to trace the entire oeuvre of publications assigned to a single individual. This is complicated by the lack of reliable author disambiguation tools. Despite advances within bibliometric databases (such as Web of Science and Scopus) and by outside sources (such as ORCID), there remain challenges in identifying the full publication history for an individual. Given the relatively low adoption of ORCID, several scholars have sought to algorithmically disambiguate authors with the same name to create profiles for individual scholars.[23] We use such an algorithm in this section (as well as in Chapter 1).[24]

The combination of author disambiguation and gender assignment algorithms, as well as a better indexing of researchers' affiliations on each of their papers, provides an opportunity to perform large-scale analysis of the mobility of publishing scientists. Recent studies utilizing a bibliometric approach have demonstrated that mobile scholars generally have a higher productivity and scholarly impact—although this varies across countries.[25] However, variation by gender remains underexplored.

The analysis in this chapter is of relatively early- to midcareer researchers: all researchers who published their first paper between 2008 and 2010 ($n = 2,528,989$) for whom a gender could be assigned based on the characteristics of their given and family names.[26] The analysis ends in 2020; therefore, this represents the first decade of these researchers' careers. While the age at which a researcher gains independence varies by field—and has increased over time—it is plausible that this time window captures mobility events between their first publication as a student and subsequent publications as a postdoctoral researcher or assistant professor (or synonymous position in a given country) and possibly into their next career stage (for example, associate professor).[27]

To consider the various forms taken by mobility, and following the typology detailed in previous work, we divide researchers into two categories: travelers and migrants.[28] Travelers are those who add new affiliations from a different country while maintaining a link with their country of origin (the country where their first paper was published). Migration means losing, at some point over the decade of analysis studied, their link with their country of origin (having at least one publication without affiliation to the country of origin).[29] Our analysis focuses only on international mobility; therefore, one could have several institutional moves within the timeframe but not be considered mobile. These results provide a novel lens on mobility and gender, examining a common cohort across countries and disciplines and relating this to the key variables discussed elsewhere in the book (namely, production, collaboration, and impact).

The Productive Nature of Mobility

Despite the policies promoting it, international mobility is relatively rare within the population of publishing scientists.[30] For the vast majority of researchers, the country in which they publish their first article is the one to which they remain affiliated. For those whose first article was published between 2008 and 2010, only 6.1% had demonstrated international mobility by the end of 2020, as per their publications. This is explained largely by attrition in scholarship: 61.3% of researchers who published their first article never published again, which de facto limits their likelihood of mobility. And the vast majority of researchers in our dataset (82.1%) have published five articles or fewer. For researchers who have authored more than one article—and are therefore more than transient authors in the research system—we observe much higher rates of mobility, as well as striking differences across gender. As the number of publications increases, so too does the probability of mobility. For example, while 7.1% of researchers with five articles are migrants (and 2.6% are travelers), this percentage increases to 22.3% for migrants (and to 5.1% for travelers) for those who have fifteen articles. This is partly a statistical artifact: given that mobility is measured through affiliations on publications, an increase in publications will inherently increase the probability of mobility. It is also a function of scientific aging and the passage of time, with possibilities of mobility increasing across a career.

Gender differences between traveling and migration are also present and interact with measures of production. In aggregate, about 1.7% of women, compared with 1.9% of men, in our cohort of young researchers are travelers—that is, they have added a country of affiliation in addition to their origin country. The gender difference is slightly higher in the case of migration (when a scholar leaves their origin country), at 4.0% for women and 4.5% for men. Mobility increases as productivity increases: among those researchers who have at least ten articles, 6.0% of women and 6.6% of men are travelers, and 21.7% of women and 26.7% of men have migrated. While the gap between men's and women's migration levels increases in absolute terms with the increase in research production, it stays stable in relative terms, with men about 20% more likely to migrate, on average, than women with the same number of articles. Men are also more likely to have a higher number of total countries with which they are affiliated—and the gender gap widens as the number of countries increases—although the overall gender difference is quite small (Figure 5.1, inset). The stability of the results demonstrates that production and mobility are intricately connected but a gender gap remains, particularly in the case of migration.

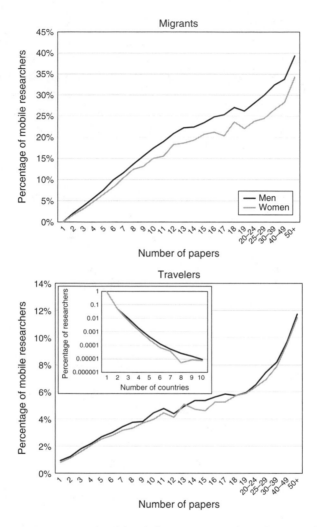

FIGURE 5.1. Proportion of mobile scholars (migrants and travelers), as a function of number of papers and gender. Inset: percentage of researchers by number of countries. Cohorts of 2008–2010, mobility for 2008–2020.

The Disciplinary Invariance of Mobility

Mobility effects vary dramatically by discipline, especially in terms of migration (Figure 5.2). Researchers are much more likely to migrate in physics than in health. This speaks to the international nature of certain disciplines—research problems studied by physicists are generally not geographically situated, while those of health sciences often are—and the requirements of mobility aligned with particular types of instrumentation (for example, astronomical observatories and particle accelerators). Merieme Chadid, for example, would not have been able to do her work without mobility. Gender differences in both migration and traveling are observed in almost every discipline. However, it is in the disciplines that have the highest mobility that we see the smallest differences in gendered mobility (physics, earth and space sciences). This is another instance of the parity paradox. While this finding demonstrates relative equality in mobility in certain disciplines, it is notable that these are disciplines with some of the largest gender gaps in participation. It may suggest, therefore, that the requirement of mobility in these disciplines serves as a barrier for participation. Parity may be a result of attrition, rather than equity: only those women who can be mobile remain active in those disciplines.[31] Disciplines for which mobility is not required, however, tend to have far higher rates of mobility for men. Mathematics, humanities, and arts, for example, do not typically require mobility, and here we observe the largest gap in gendered mobility.[32]

Those who have traveled but never migrated occupy a smaller proportion of the scientific workforce. Here, too, we see gender differences: artists and social scientists tend to be more likely to gather affiliations without migrating, but men dominate in this type of mobility. Reinforcing what we have observed in the funding chapter (Chapter 3), it is in those disciplines where women are in the minority—such as physics, earth and

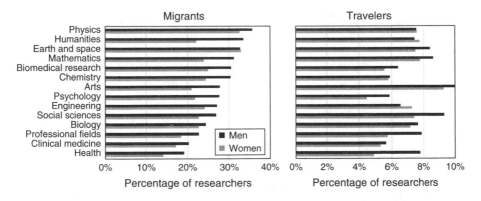

FIGURE 5.2. Proportion of mobile scholars (migrants and travelers), by discipline and gender, for researchers of the 2008–2010 cohorts who have published at least ten papers, mobility for 2008–2020.

space sciences, and engineering—that they tend to have higher rates of mobility.

Does It Matter Where One Starts Their Career?

Country of origin is a major factor in gendered mobility. Even among the most productive countries, we observe strong gendered differences in mobility (Table 5.1). In Israel, for example, men are more than twice as mobile as women, with more than 10% of Israeli men leaving the country. Sweden demonstrates similar rates, with 10% of the men migrating, compared with 6.8% of women. The only two countries where women are close to equal or have higher mobility than men are Japan and South Korea, nations where women are highly underrepresented in authorship. While these two countries also are among those with a lower percentage of women authorships, we did not observe any large-scale relationship between the percentage of women authorships in a country and their probability for mobility. Motivations for movement may be a combination of scientific resources and cultural perspectives in a scholar's home country. This suggests that policies may need to consider the implications of both short- and long-term stays, particularly given personal constraints that may allow traveling but restrict migration.

If we examine only migration (the most common form of mobility), we see that, in most Western countries, as well as in most countries with higher economic development (except Japan and South Korea, as mentioned earlier), men are more likely to exhibit mobility than their women colleagues (Figure 5.3). In the United States, for instance, while 3.7% of men from the 2008 cohort are mobile, only 2.8% of women researchers are mobile. Although these two percentages may seem low, relative difference between the two genders remains high (32%) and consistent when accounting for differences in production. Lower rates of mobility for women are also observed in the Middle East and northern Africa, reinforcing previous results on the higher mobility of men in these countries.[33]

Although the brain drain metaphor has been replaced by the notion of "circulation," there are still those who tend to win and lose in the mobility game. North America tends to gain migrant scholars, whereas Europe and Asia tend to lose them.[34] This is complicated, however, when distinguishing between migrants and travelers: countries like Saudi Arabia, for example, have high net gains from travelers, largely due to programs

TABLE 5.1. Percentage for men and women who are mobile (migrants and travelers), for the thirty countries with the highest scientific production; 2008–2010 cohorts, mobility for 2008–2020.

Country	Women			Men			Men/women difference in combined mobility
	Migrants	Travelers	Combined mobility	Migrants	Travelers	Combined mobility	
Israel	5.3	1.1	6.3	10.9	1.9	12.8	202.1
Czechia	4.3	2.6	6.9	7.5	3.6	11.2	162.3
Iran	5.5	1.7	7.2	8.5	2.6	11.1	154.2
India	4.7	0.9	5.6	7.2	1.1	8.3	149.1
Canada	5.3	1.5	6.9	7.5	2.4	9.9	143.8
Finland	5.4	3.2	8.6	8.6	3.7	12.4	142.8
Sweden	6.8	3.5	10.3	10.0	4.6	14.5	141.5
Poland	2.8	1.9	4.7	4.0	2.7	6.7	141.2
France	7.8	2.7	10.4	11.4	3.3	14.7	140.5
Australia	6.2	2.0	8.2	8.4	3.1	11.6	140.4
Germany	6.4	2.1	8.6	9.1	2.5	11.6	135.6
United States	2.8	1.2	3.9	3.7	1.6	5.3	134.1
Belgium	8.6	3.1	11.7	11.4	4.2	15.6	132.8
Brazil	2.9	1.9	4.9	4.0	2.4	6.4	131.3
Austria	8.6	3.1	11.7	11.8	3.4	15.2	130.5
Mexico	4.9	1.6	6.4	6.5	1.9	8.4	129.6
Denmark	7.0	3.8	10.8	9.9	3.8	13.8	128.0
Italy	6.9	2.3	9.1	9.2	2.5	11.7	127.8
Russia	2.6	1.2	3.8	3.4	1.5	4.9	127.8
Netherlands	9.7	3.3	13.0	12.5	4.0	16.4	126.4
United Kingdom	7.8	2.1	9.9	10.0	2.4	12.4	125.0
Switzerland	10.9	3.3	14.1	13.9	3.6	17.5	124.1
Turkey	2.7	1.5	4.2	3.6	1.5	5.1	122.8
Taiwan	0.9	0.5	1.4	0.9	0.7	1.6	117.7
Singapore	7.9	2.6	10.4	9.4	2.7	12.1	115.7
Spain	6.4	2.4	8.8	7.9	2.2	10.1	115.1
Portugal	6.7	4.2	11.0	8.5	3.9	12.4	113.3
China	1.1	1.4	2.5	1.2	1.6	2.8	109.9
Japan	2.2	1.2	3.4	2.2	1.1	3.3	99.5
South Korea	2.1	0.8	2.8	1.5	0.7	2.1	75.7

Women more likely ▬▬▬▬▬ Men more likely

FIGURE 5.3. Difference between the percentage of men who are migrants and the percentage of women who are migrants, by country; 2008–2010 cohorts, mobility for 2008–2020. The lighter the country, the more likely women from that country are to migrate to another country, compared with men; the darker the country, the more likely men from that country are to migrate to another country, compared with women.

that encourage association through affiliation.[35] If we examine all mobility types in the aggregate, we observe gendered differences in directional mobility. Table 5.2 displays the difference between women's and men's mobility to particular regions, as a function of the region in which they published their first article. Those below 100% (lighter shade) indicate that women are more likely to move to that destination (column); values over 100% indicate that men are more like to move from a particular region (row) to the destination. As shown, certain regions of the world are more likely destinations for women, regardless of origin: those include all European countries, Oceania, and the Americas. Those destination countries likely to retain or attract men are primarily Asian and African countries. There are, however, some differences observed: men who publish first in northern Africa are more likely than women to go to northern Europe, whereas women from the same region go to southern Asia.

There are a few important limitations to these data. Origin of first publication can only be loosely interpreted as a proxy for nationality, given that for many students studying abroad, their first publication is in a foreign country. This was observed in a previous analysis that demonstrated that the large degree of flow of researchers from North America (origin) to Asia (destination) was not a case of North American emigration but rather a reflection of the high number of Chinese-born students studying in the United States.[36] This provides a nuance to interpreting Table 5.2—the high

TABLE 5.2. Difference between women's and men's likelihood to move to a region of destination, as a function of the region of origin, for mobile researchers (travelers and migrants); 2008–2010 cohorts, mobility for 2008–2020 (percent). Lighter cells mean that women from that region of origin are more likely to move to that region of destination; darker cells mean that men from that region of origin are more likely to move to that region of destination.

Region of origin	Region of destination												
	Northern Europe	Western Europe	Southern Europe	Australia and New Zealand	North America	Southeast Asia	Eastern Europe	Sub-Saharan Africa	Latin America and the Caribbean	East Asia	West Asia	Northern Africa	Southern Asia
Northern Europe	82	95	90	83	96	95	91	94	108	187	154	200	198
Western Europe	87	101	84	83	101	100	71	94	94	164	166	93	199
Southern Europe	91	99	87	109	97	146	104	184	111	158	161	125	312
Australia and New Zealand	81	96	106	92	89	88	149	86	86	160	185	88	167
North America	86	89	66	76	88	86	84	70	78	131	140	151	158
Southeastern Asia	72	68	98	80	89	88	139	143	97	105	310	285	194
Eastern Europe	101	104	76	64	102	57	90	129	135	109	174	270	418
Sub-Saharan Africa	86	89	113	93	78	108	183	145	128	119	127	107	146
Latin America and the Caribbean	101	98	100	86	94	151	127	86	109	198	121	522	308
East Asia	104	106	107	111	96	87	85	102	96	95	201	139	179
West Asia	72	83	97	124	88	233	70	131	134	215	169	127	200
Northern Africa	106	57	78	126	151	135	225	130	361	302	130	113	84
Southern Asia	96	93	80	83	86	116	90	123	95	142	150	335	156

dominance of men landing in Asia can be interpreted as Asian men staying within the region at higher rates; it can also be an indication that women are less likely to return when they leave. Understanding the directionality of mobility is key for insights on both the distances of mobility and the regions that provide hospitable climates for women's work.

Countries also vary substantially by their degree of international mobility. US international mobility, for example, is notoriously lower than that of most other countries.[37] One explanation is an artifact of size: that is, the number and concentration of academic institutions in the United States allows for scholars to receive the benefits of mobility without the costs of moving (including, for example, transitioning across different economic and language barriers). Movement, however, is not completely open; as computer scientist Aaron Clauset and colleagues have shown, institutional mobility within the United States follows a "steeply hierarchical structure that reflects profound social inequality."[38] European researchers face a different dilemma: whereas movement may occur more easily across borders due to agreements within the European Union, there remain cultural and language barriers. These barriers in language are particularly present for highly localized domains—such as education, social work, and other "care"-related disciplines (see Chapter 1)—where women are disproportionately represented.[39]

Interplay with Collaboration and Impact

Mobility does not happen in a void: it is, in many cases, the result of international collaboration, and it can be associated with other indicators of research impact. Figure 5.4 presents how scientific impact (top panel) and international collaboration (bottom panel) evolve with different types of mobility, as well as across levels of research productivity. Figure 5.4 also shows how scientific impact increases with productivity, a finding that replicates previous research.[40] However, it also demonstrates that nonmobile researchers have lower impact than mobile researchers and that, among mobile researchers, migrants have higher impact than travelers. There are also no sizable differences observed by gender. This suggests that, when mobile, men and women both benefit similarly in terms of citations from the international exposure associated with mobility. Nevertheless, given that women are less mobile, they are less likely to benefit from mobility collectively.

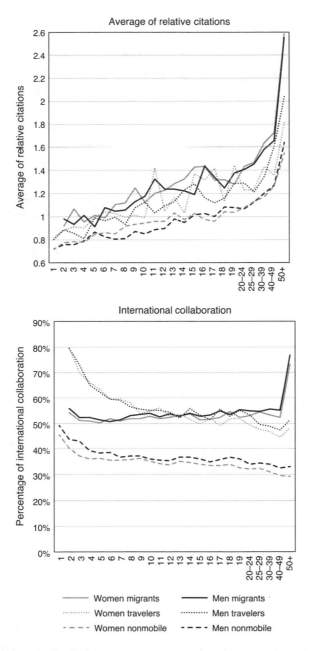

FIGURE 5.4. Average of relative citations (top panel) and mean international collaboration rates (bottom panel) of mobile (travelers and migrants) and nonmobile researchers, by research productivity; 2008–2010 cohorts, mobility for 2008–2020.

In terms of international collaboration, nonmobile researchers are much less likely to collaborate internationally than their mobile colleagues, and such collaboration decreases as their research productivity increases. However, quite strikingly, men who are not mobile have higher international collaboration rates than nonmobile women. This suggests that even when they are not mobile, men are more likely to benefit from the extra exposure associated with international collaboration—although that does not appear to translate into a particular citation advantage (Figure 5.4, top panel). Travelers are decreasingly collaborative with international colleagues as their research production increases, while international collaboration slowly increases with numbers of articles for migrants—until those researchers with fifty publications or more, for which this percentage is the highest. However, no sizable gender differences in international collaboration are observed for both migrants and travelers. The interplay between resource and reward is clear: women are less likely to be active in international collaboration, are less likely to be mobile, and thereby suffer consequences in terms of citations.

The Global Researcher

Science and more broadly academia are gendered at their very access points and also in specific practices of international collaborations and conferences, including resources, opportunities, and the concept of safety abroad. International science is organized around a narrow image of the hegemonic male scientist who has resources, can take risks, and has the power and status in his university and among his peers to obtain institutional and grant support. He has no trouble asking for that support, maintains academic and social networks that can introduce him to colleagues abroad, and can draw on former students and postdocs around the world. This stereotypical image is self-perpetuating: Because he looks like the ideal global scientist, he becomes even more so.[41]

Mobility is deeply intertwined with the other bibliometric variables discussed in this book. As sociologist Kathrin Zippel suggests in the foregoing quote, mobility both is dependent on and contributes to production, collaboration, funding, and impact. These cumulative disadvantages are reinforcing. For example, if one does not have the resources necessary for circular mobility—such as attending conferences—longer-term and more influential mobility events are less likely. International visibility, in turn, is a strong mechanism for predicting international collaboration: in a study of one hundred National Science Foundation–funded principal investigators, Zippel found that there was a gender gap in the ways in which collaborations were formed, with men more likely to draw collaborators from international experiences.[42] Even short-term mobility, therefore, is imperative for women to have similar access as men to the global scientific community. As we have shown in previous chapters, women have lower rates of international collaboration, and this has consequences for scientific impact. All these events are dependent on proper resources: mobility comes at a cost, and women are less likely to have the funding necessary for travel, particularly if they have added personal costs (such as childcare). The disadvantages observed throughout this book—in production, collaboration, labor, funding, and mobility—leads to reinforcing disparities at the intersection.

Our results demonstrate that, on average, women are less likely to have circular mobility (traveling) or directional mobility (migration). We find a relationship with both productivity and impact, with greater degrees of mobility linked to both higher numbers of articles and increased citations. These trends hold across disciplines and countries, with a few important exceptions. We find that women tend to be as mobile as or more mobile than men in fields that have required mobility (such as earth and space sciences) rather than in fields in which mobility is optional (and more of a luxury) (such as humanities or mathematics). We also find higher rates of mobility for women in countries where resources are scarce—several countries in sub-Saharan Africa—and where women scientists are in the extreme minority (Japan). On the other hand, women were less likely to be mobile in North African and Middle Eastern countries, which, in many cases, do not grant equal rights to women.[43] Gendered issues of mobility are deeply connected to other sociopolitical characteristics of the scholars and the nations from which they originate. Women's lower rates of mobility observed consistently throughout this chapter reflect these issues.

Scientific resources are not distributed equally around the world. Large instruments, such as astronomical observatories or the Large Hadron Collider, create physical anchors that draw scholars to specific locations. Facilities such as medical labs and hospitals are dependent on certain levels of governmental investment and political security in a country.

Disciplinary results reflect this: mobility is essential for earth and space sciences but not as prevalent in health and social sciences. The prestige and concentration of personnel in a location is another factor: having many experts in a given area creates a hub of activity that attracts other collaborators and colleagues, while the lack of available positions may lead researchers to leave the country. Simply put, there are not equal opportunities to advance science in every location in the world. Therefore, an optimal system would be one where talented scholars have access to the instruments, facilities, and collaborators they need to do the best science. Unfortunately, this is not necessarily the case, nor is it the case equally for all sectors of the population.

Scientific mobility at the individual level is often linked to mobility in other aspects of life: those from families who have traveled or lived abroad will have a higher propensity for professional mobility. Individuals from disadvantaged backgrounds or who have otherwise limited social mobility may be less mobile. For example, in the United States, students of color are underrepresented in study-abroad programs, with consequences for later mobility.[44] Early experiences in mobility for "majority" students may have an effect in further propensity for scientific mobility. However, although women participate at higher rates in international study-abroad programs and earn degrees abroad more than men, the trend reverses at the postdoctoral level, with higher rates of men demonstrating mobility at that career stage—at least within the European Union.[45] This disengages women from international networks that can bring mobility and collaboration opportunities and, in turn, lead to greater academic visibility and prestige. Similarly, as postdoctoral fellowships become necessary and as research stays abroad become the norm in certain fields and countries, the potential for women to progress in academic careers is constrained.[46]

Countries also have a role in fostering mobility: political climate and immigration policies affect the circulation of talented scholars around the world.[47] The United States, for instance, is highly dependent on the educational investments of other countries, as a sizable percentage of its research workforce was trained abroad. Trends toward isolationism, however, can be seen across the world—from Brexit to Executive Order 13769—which threatens scientific mobility.[48] Such isolationism could increase the gender gap in mobility: as it becomes more difficult to move from one country to another, only researchers with fewer domestic constraints can travel, and those are more likely to be men.[49] We may also observe gendered differences as science emerges from the consequences of the pandemic, where the interplay between state policies and family dynamics may have dramatic consequences for global science.

Marriage is another barrier that particularly disadvantages women: women are disproportionately "tied movers"—that is, intimately partnered with another highly skilled worker.[50] Given the expectation of mobility in scientific careers, this often means that mobility comes at the expense of women's career trajectories: in the case of dual scientific careers, men's careers are prioritized and mobility serves to displace women scientists, relegating them to lower-tier career options.[51] For women with children, there are additional barriers to mobility, particularly when they do not have a large support network at home in the form of an extended family or partner. Even with this, women academics often report high levels of guilt associated with their travel.[52] When mobility implies choosing between their family and their career, women face greater pressures than men and their mobility is consequently more often limited.[53] A striking example of gendered differences in family mobility can be seen in the Humboldt research fellows in Germany: in the 1980s, 60% of women visiting scholars came alone, whereas 60% of the men relocated with their families.[54] Mobility policies, therefore, must consider the dynamics of the work-family conflict.[55]

One must also note the safety risks that women face in mobility compared with men, which further limit their access to certain places. Fieldwork, for example, is rife with safety concerns in countries where women cannot move as fluidly in society. Take, for example, the experience of Jane Willenbring, who, as a graduate student in geology at Boston University, took the long trip to Antarctica with a small research team, including her adviser, David Marchant. For scientists, the trip is one that involves transportation on a military aircraft, where passengers are itemized as "self-loading cargo" on the manifest. Before boarding the last plane in Christchurch, New Zealand, the self-loading cargo must dress for the temperature shock they are about to experience. On board, there are no bathrooms, just five five-gallon buckets tucked in the rear of the aircraft. The flight is nine hours to McMurdo Station and there is no meal service.[56] For many travelers, the entire trip can take a week with all flights and delays, and the commitment on land is often many months. Willenbring kept traveling beyond the main station—she was dropped on location with her small group of compatriots and spent weeks with only radio contact with McMurdo. During her time there, she reported experiencing physical assault and sexual harassment from her faculty adviser. Willenbring waited to report the incident until she was tenured and, consequentially, less vulnerable professionally. Her reporting was corroborated by other women with similar experiences on their trips with Marchant, who was later fired by Boston University after an investigation of the allegations.[57]

This is in no way an isolated experience: a majority of women faculty have experienced sexual harassment in the academic workplace.[58] However, the isolation of the research environment allows for power dynamics to be exploited in serious ways. This has been demonstrated in other areas of study that require field research; for example, 26% of women archaeologists reported experiencing sexual assault and 71% sexual harassment while conducting field research.[59] It is imperative, therefore, that safety be taken into consideration in individualized mobility experiences (such as fieldwork) as well as collective ones (such as conferences). Conference organizations that are attentive to equity and include codes of conduct may help improve safety for all scholars. Another potential solution is to increase the opportunities for women to host and provide mobility events for other women. Homophily effects, demonstrated in other chapters, are also present here. For short-term stays—or circulation mobility—women are more likely to host other women.[60] Therefore, increasing the number of senior women in a field is also likely to have positive effects on women's mobility opportunities and possibly increase safety.

Gender equity is a multifaceted and systemic phenomenon; all the variables we have discussed in this book, including mobility, are interdependent and reinforcing. Using mobility as an indicator of academic success requires that institutions and organizations provide equal opportunities for all members of the scientific community to engage in these knowledge-transfer activities. This includes providing funding to cover costs for families to travel and to provide resources for women to host international workshops at their home institutions when traveling is not possible. Furthermore, it is debatable whether mobility should be used as an indicator of academic success at all, as mobility is a luxury that many cannot afford. This capital also gives way to other forms of capital (such as citations), as demonstrated. We move, therefore, to a deeper examination of citations, the most often used indicator of scientific impact in science.

Chapter 6

Scientific Impact

Very few scientists win a Nobel Prize. Even fewer win it twice.[1] Marie Curie was one of the select few to receive this high recognition of scientific impact. She has captured the imagination of scientists around the world. As a successful woman scientist in the early twentieth century, Curie was something of a curiosity and, as such, has been critiqued and venerated by scientists and the lay public alike. Feminists adopted Curie for their agenda, and she was paraded around the United States in a much-publicized tour in 1921. However, the tour somewhat backfired for those hoping to promote the cause of women in science.[2] Rather than empowering women, the case of Curie had the effect of magnifying the inferiority complexes of women: by using Curie as an exemplar, no woman scientist was able to compare. The "Madame Curie Complex," as historian Julie Des Jardins has termed it, "empowered and stigmatized women, liberated and constrained them, often at the same time."[3] Evidence of this complex can be found in the derivative nicknames of successful contemporaries and successors. These names are intended to be laudatory but often place their bearers in the shadows.

Lise Meitner (1878–1968) was one such woman. She was fondly called "our Marie Curie" by Albert Einstein, but she received no mentorship from Curie.[4] In fact, Meitner applied to work with Curie early in her career but was rejected.[5] Her career might have taken a vastly different turn had she landed in Curie's lab. Meitner was born into an upper-middle-class Jewish family in 1878 in Vienna and persisted through various educational institutions, despite explicit discrimination against women. She was privately educated in physics, graduated from *gymnasium* (preparatory high school) in 1901, and, in 1905, became only the second woman to obtain a doctoral degree in physics.[6] She moved to Berlin in 1907 hoping to attend lectures at the University of Berlin, unaware that women were barred from Prussian academies. She gained permission from Max Planck to attend his lectures, an opportunity not typically awarded to women. Planck's view on women in science was clear: "I must hold fast to the idea that such a case must always be considered an exception, and in particular that it would be a great mistake to establish special institutions to induce women into academic study, at least not into pure scientific research . . . In general it can not be emphasized strongly enough that Nature itself has designated for woman her vocation as mother and housewife, and that under no circumstances can natural laws be ignored without grave damage."[7]

Despite this ideology, Planck served as an essential ally for Meitner. One of his most influential acts was advocating for her admittance to the Chemistry Institute, a building that women were strictly forbidden from entering. After her arrival in Berlin, Meitner met another young scientist, Otto Hahn. To facilitate their burgeoning collaborative relationship, she was allowed to frequent the building where he worked as long as she entered through the basement and did not disturb the students in the Chemistry Department. She complied, and thus began a three-decade collaboration that resulted in a Nobel Prize. But not for Meitner.[8]

In 1912, Meitner and Hahn moved their lab to the newly founded Kaiser-Wilhelm-Gesellschaft in Berlin, where Meitner worked without salary as a "guest" of Hahn, who was given both a salary and the title of professor. This was in keeping with the practices of other institutions: academic titles were strictly unavailable to women in Germany in the early twentieth century. Such policies, however, did not govern the privately funded Kaiser-Wilhelm-Gesellschaft. With the instigation of Planck, Meitner received the paid position of assistant by the end of 1912—the first woman with such a position in Berlin—and ascended through the ranks, eventually heading her own section for physics at the institute and obtaining the title of professor in 1926, the first woman to do so in Germany. These advances halted with the 1933 civil service law in Germany, which introduced several provisions to limit the advancement of women, "non-Aryans," and other "undesirables" in civil services. The private funding of Meitner's institute provided some protections, but the social situation in Germany quickly became untenable for a Jewish woman scientist.[9] In 1937, the Kaiser-Wilhelm-Gesellschaft was reorganized and placed under governmental control.

In June 1938, Meitner learned that she would be dismissed but forbidden to emigrate, for fear that her intelligence would be used against Germany. She fled to Stockholm by way of Holland, arrived penniless, and entered a relatively hostile working environment.[10] Despite this, she continued her research and correspondence with Hahn. In 1938, they finalized the theoretical and experimental components that led to the discovery of nuclear fission. The political climate prevented joint publications, but their shared contributions were noted by their joint nomination for a Nobel Prize in 1939. In 1945, the Nobel Prize in Chemistry was given to Hahn. Meitner received no prize, and her contribution was reduced to that of a *Mitarbeiterin,* despite large protestations from the scientific community.[11] Once the records of the Nobel Committee's proceedings became public, her case was reviewed. The summative statement of the analysis declares that "Meitner's exclusion from the chemistry award may well be summarized as a mixture of disciplinary bias, political obtuseness, ignorance, and haste."[12] Scientific impact was not part of the rationale for excluding Meitner.

Another woman physicist whose impact was underrecognized is Chien-Shiung Wu (1912–1997). Like Meitner, she received a diminutive Curie title, this time indicative of her race: "the Chinese Marie Curie."[13] Wu was born in Liuhe, a town in the Jiangsu province of China. She received her elementary education at a school for girls founded by her father, attended a boarding school for her secondary education, and then studied mathematics and physics at National Central University in Nanjing. Unlike Meitner, Wu's gender was not an impediment to her education. Following graduation, she received a research position at the Institute of Physics of the Academia Sinica. Her adviser at the institute was a graduate of the University of Michigan and encouraged her to follow suit. She left China for the United States in 1936, never to see her parents again.

Wu's boat landed in San Francisco, and she redirected to the University of California, Berkeley, after learning that women were not allowed in the front doors at the University of Michigan. She noted the extreme oppression of women in America, which she found to contrast with her experience in China. In an interview in 1963, she noted, "US society and families unfortunately believe that science and some other fields are exclusively men's turf . . . It is different in China . . . In the 1930s, Chinese society realized that we had to deploy all resources—collective talents of both men and women—if we want to catch up with the West . . . The West is ahead of China in science and technology, but not necessarily in the effective utilization of human talents."[14]

However, it was her race rather than gender that was a barrier at Berkeley, where Asian students were funded at lower levels than their white counterparts. She moved to the California Institute of Technology, where she received a full scholarship, completed her PhD in 1940, and worked as a postdoctoral fellow in the Radiation Laboratory. Wu took a brief position at Smith College, a private women's college, and then the naval academy at Princeton. In 1944, she joined the Manhattan Project's Substitute Alloy Materials Laboratories at Columbia University by the invitation of her former professor, Robert Oppenheimer.[15] She was the only Chinese scholar and one of the only women among senior researchers to join the Manhattan Project.

After the war, Wu received a full-time position at Columbia, although she was untenured and underpaid compared with her men colleagues.[16] In 1956, she was approached by two colleagues, Tsung-Dao Lee and Chen Ning Yang, who had proposed a theory regarding the invalidity of the conservation of parity. They asked Wu to design and execute an experiment to test their hypothesis.[17] The "Wu experiment," as it would be called, proved their theory and served as a major contribution to particle physics. The three scholars—Wu, Lee, and Yang—were featured on the covers of the *New York Times, Newsweek,* and *Time.* The Nobel Prize was an obvious conclusion but only two were invited to Stockholm in 1957. Wu was not among them, despite protestations from Oppenheimer, Lee, and Yang. There were many firsts for Wu: she was the first woman to receive an honorary doctorate from Princeton (1958), the first woman president of the American Physical Society (1975), and the inaugural winner of the Wolf Prize (1978).[18] But she was not to win a Nobel.

In 2018, Donna Strickland was the first woman to be awarded the Nobel Prize in Physics in fifty-five years. Born in Guelph, Ontario, Canada, she received her undergraduate degree at McMaster University and performed her graduate work at the University of Rochester in the United States. It was during her graduate time that she published her prize-winning work, together with her doctoral adviser, Gérard Mourou. After graduation, Strickland worked at the National Research Council of Canada, the Lawrence Livermore National Laboratory, and Princeton University's Advanced Technology Center for Photonic and Opto-electronic

Materials. When she joined the University of Waterloo as an assistant professor in 1997, she was the institution's first full-time woman professor in physics. When she was awarded the Nobel, her gender again received notable mention in the press. Strickland noted surprise at this, stating, "I don't see myself as a woman in science. I see myself as a scientist. I didn't think that would be the big story. I thought the big story would be the science."[19]

As the press began to scrutinize Strickland's history and academic achievements, other notable items began to emerge. One was that, at the time she was awarded the Nobel Prize, she was still an associate professor, having never applied to be full professor. The story gained additional traction due to her missing Wikipedia page. A few months before she was awarded the Nobel Prize, a new volunteer Wikipedia editor submitted a draft article on Strickland. An editor rejected the submitted entry, saying that Strickland did not meet Wikipedia's "notability requirement."[20] Apparently, even Nobel-merit research—which can be considered the pinnacle of scientific impact—did not make a woman scientist notable.

Measuring Scientific Impact

Awarded for research that had "the greatest benefit to humankind," Nobel Prizes constitute the most extreme recognition of scientific impact.[21] Between 1901—the year the Nobel Prize was first awarded—and 2021, only 714 distinct individuals have won one of the four scientific prizes.[22] Only 25 of the medalists are women, representing 3.5% of all awardees.[23] Given this singularity, Nobel Prizes are ill-suited for wide-scale evaluation of researchers. Therefore, we rely on more universal indicators of research impact—citations. As noted in the introduction, authorship serves as a central anchor in the reputation cycle of academic work. By extension, citations function as a form of currency exchange: authorship denotes production of a good, and a citation signals that the good was consumed. The higher the demand for the work, the higher its value and, by extension, the authors, institutions, and countries with which it is associated. This assumption, of course, requires both contextualization and interrogation.

The Science Citation Index was created in the second half of the twentieth century as a device for information retrieval, creating a web of knowledge that would allow researchers to traverse the increasingly complex and voluminous space of scientific documents. This web was formed through citation linkages: that is, the link between a citing and cited document.[24] By the 1970s, and following the leadership of information scientist Francis Narin, evaluators had begun to use the number of citations of a given work as an indicator of its scientific impact.[25]

The shift to electronic databases in the 1990s further expanded the use of citation indexes, not only for information scientists but for the wider scientific community. Citation indexes became the dominant source of data for large-scale scientific evaluation, with citations as the main indicator of scientific impact.[26] Citation indicators were refined and reified over time and became a cornerstone of research evaluation systems.[27] In particular, the manifestation of citations in national evaluation schemes and promotion and tenure criteria solidified the importance of these indicators for symbolic capital and esteem.[28] Given this importance, any systemic bias in this indicator is likely to have consequences for scientific trajectories.

Understanding citation motivation is key to unlocking potential biases in citation indicators. Several factors can influence why scholars cite (and why they do not).[29] Two main perspectives are represented among citation theories: the normative and constructivist perspectives.[30] Following a normative framework, authors give credit to other authors for their past achievements through cumulative citation practices. In this way, references provide "pellets of peer recognition" for past knowledge claims, justifying the use of citation as a form of academic currency.[31] Socioconstructivists, however, argue that references are rhetorical devices that aim to persuade both referees and readers alike of the validity of the knowledge claims that are contained in a citing paper.[32] Proponents of this perspective tend to argue that these decisions to cite are based on a nonscientific or nonmeritocratic rationale, thereby undermining the use of citations as impact indicators.

We argue for a third perspective that draws on both theories, what could be called the utility perspective: citations are simply an indication of the *use* of the cited work in the construction of the citing work. Our perspective remains agnostic on the citer's evaluation of the merit of the given work. For example, although negative citations account for a relatively trivial percentage of all cited references, we would argue that their inclusion does not negate citation indicators but is a valid representation of utility.[33] Engaging with a previous piece of research, even if negatively, marks a certain level of interaction and provides evidence of cumulative knowledge production. The same is true for a work cited for persuasion: in evoking a work, the citing author is demonstrating intellectual influence. Therefore, whether motivated by scientific or social reasons, positively or negatively, citation links are a demonstration of the utility, and thereby *impact,* of scientific work.

The danger, however, comes in the equation of citations with quality and the implications for what is not cited or is lowly cited. While quality may be a latent property of citedness, it is heavily outweighed by several other orthogonal factors, largely related to visibility. For instance, the journal in which a paper is published has been shown to affect its citation rates, independently of its intrinsic quality.[34] Journals that are more visible to the international community are likely to attract more citations: this visibility may be a function of the prestige of the publisher or scientific society associated with the journal, inclusion in a citation index, or availability via open access.[35] Document-level characteristics are also an important function: articles in certain disciplines or on specific topics are more prestigious, as are those written in the current lingua franca of science—English—which can be read by researchers from a broader set of countries and generate higher citation counts. Geography and proximity are also important: people are more likely to cite authors who are within their same university, region, and country.[36] This, too, is an effect of visibility. To attract more international citations, one must be mobile: presenting at scientific conferences, engaging in personal mobility exchanges, and, increasingly, being present on various online forums. These factors of visibility do not distribute equally across the population (see Chapter 5), thereby introducing potential forms of inequity into citation indicators. That which is visible is more likely to be used, and increased use, in turn, increases visibility.

Given the importance of citations as the currency of the scientific ecosystem, several studies have sought to examine potential gender disparities in impact. Analyses have focused on a range of disciplines—astronomy, demographics, economics, engineering, management, natural language processing, neuroscience, and physics, among many others—and specific nations, with few studies analyzing trends at the macro level.[37] The overwhelming majority of studies have found that women's work was less cited than that of men, with a few exceptions.[38] Extending and building on this previous work, we seek to examine the different rates of citations of papers to which men and women contributed across disciplines and countries. To address field differences in citation practices, most of the results in this chapter are based on field-normalized citation rates, which normalize each article by the average citation rate of other articles published in the same specialty during the same year (average of relative citations).[39] The same method is applied to journal impact factors (average of relative impact factors). Such average measures have been shown to be imperfect; therefore, we also provide indicators of the two extremes of the citation distribution: uncited papers and highly cited publications.[40] Given the influence of collaboration practices on the impact of scholarly papers, as well as gender differences in collaborative practices (see Chapter 2), our analysis distinguishes gender differences in scholarly impact by the type of research collaboration: single-authored publications, publications in national collaboration only, and publications in international collaboration.[41]

What Is an Article Written by a Woman?

In 2017, Elsevier—the world's largest publisher and owner of the Scopus bibliometric database—released a report examining the evolution of women's contributions to science over the course of the previous two decades. Some of the results were encouraging: in each of the countries analyzed, the proportion of women active in research increased between 1996–2000 and 2011–2015, reaching parity in countries such as Brazil and Portugal.[42] The report also presented data on citations, concluding that parity had been achieved: it stated that, globally, women's papers were cited at nearly equal rates to men's and exceeded men's rates in countries such as the United States. Unfortunately, it was not progress but rather a methodological artifact behind this surprising result.

The analysis hinges on a deceptively simple question: What is an article written by a woman? This question is easily answered in a world of single authorship but is complicated in a collaborative environment. For example, if we look at all articles in Web of Science from 2008 to 2020, we find that most articles—more than 53%—have at least one woman and one man on the byline (Figure 6.1). More than a third of all scholarly articles have only men as authors, whereas 10% have only women. These asymmetries have strong implications when calculating scientific impact. Articles written by gender-diverse teams have the highest citation rate (1.08). Those written exclusively by women are cited less frequently than those written exclusively by men (0.83 versus 0.99), and this gap is

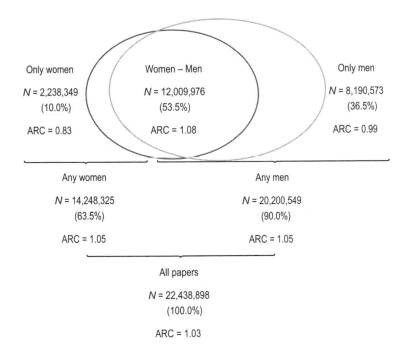

Only women

N = 2,238,349
(10.0%)

ARC = 0.83

Women – Men

N = 12,009,976
(53.5%)

ARC = 1.08

Only men

N = 8,190,573
(36.5%)

ARC = 0.99

Any women

N = 14,248,325
(63.5%)

ARC = 1.05

Any men

N = 20,200,549
(90.0%)

ARC = 1.05

All papers

N = 22,438,898
(100.0%)

ARC = 1.03

FIGURE 6.1. Number of articles and average of relative citations (ARC) of various cases associated with the assignment of an article to a woman or a man, 2008–2020.

the largest observed. However, when considering articles with at least one man (90.0% of articles) and those with at least one woman (63.5%), scientific impact is identical (1.05).

These results demonstrate the problem underlying the Elsevier report: it relies on a form of full counting, in which any articles with at least one woman as author is considered a woman's article, and any article with at least one man as author is considered a man's article.[43] Consequently, there is high overlap between the sets of men's and women's articles with the overwhelming majority of articles to which women contribute also having men among their authors. If we note dominant authorship positions (first and last authors), we find results that reinforce previous research on gender disparities in scholarly impact: fewer citations are received for articles with women in dominant authorship positions—as first authors (0.98 for women versus 1.06 for men) and as last authors (0.99 versus 1.07).[44] This demonstrates both the importance and the complexity of disentangling authorship when considering gender disparities in impact and why authorship positions and collaboration are important for understanding how gender and citations are related.[45]

Collaboration Increases Scientific Impact

Collaboration plays an important role in the visibility of scholarly articles: articles that are coauthored are more frequently cited, even when self-citations are excluded.[46] However, the relationship between the gender of lead authors and citation rates changes as a function of the type of collaboration (Table 6.1). When only single-authored articles—which obtain sizably lower citation rates—are considered, there are virtually no differences in men's and women's citation rates (0.84 versus 0.85). Disparities appear when examining national and international collaboration, using dominant authorship positions. In the case of national collaboration, we observe, for the entire period combined, roughly a 5% gap in citation impact, when considering both first and last authorship. This gap is slightly decreasing over time when last authorship is considered. The gender gap

TABLE 6.1. Average of relative citations of men- and women-led papers, by collaboration level and author position, by year, 2008–2020.

| | No collaboration | | National collaboration | | | | International collaboration | | | |
| | | | First | | Last | | First | | Last | |
Year	W	M	W	M	W	M	W	M	W	M
2008	0.88	0.95	0.98	1.04	0.95	1.04	1.29	1.45	1.27	1.40
2009	0.87	0.92	0.97	1.03	0.95	1.03	1.27	1.45	1.34	1.38
2010	0.87	0.88	0.96	1.01	0.94	1.01	1.30	1.46	1.31	1.38
2011	0.87	0.86	0.95	1.01	0.93	1.00	1.28	1.46	1.28	1.40
2012	0.84	0.86	0.95	0.99	0.93	0.99	1.29	1.47	1.29	1.41
2013	0.81	0.83	0.94	0.99	0.93	0.98	1.27	1.44	1.28	1.38
2014	0.83	0.83	0.93	0.98	0.92	0.97	1.26	1.42	1.26	1.37
2015	0.83	0.81	0.92	0.96	0.92	0.96	1.26	1.43	1.25	1.36
2016	0.80	0.78	0.91	0.95	0.91	0.94	1.25	1.41	1.25	1.36
2017	0.78	0.76	0.90	0.95	0.90	0.94	1.24	1.40	1.23	1.34
2018	0.76	0.76	0.90	0.94	0.90	0.94	1.25	1.39	1.24	1.33
2019	0.75	0.74	0.90	0.94	0.89	0.93	1.24	1.41	1.23	1.34
2020	0.78	0.75	0.87	0.94	0.84	0.93	1.18	1.43	1.17	1.35
All	0.84	0.85	0.93	0.98	0.92	0.97	1.25	1.43	1.25	1.37

in citations is, however, much greater in cases of international collaboration, with men-led papers seeing 14% higher citation rates than papers with women as first authors, and 10% higher citations in the case of last authors. In both cases, the gap persists throughout the period; there does not seem to be any improvement in the situation over time. Women are, therefore, at a compound disadvantage: they are less likely to collaborate internationally (see Chapter 2), and when they do, their work is cited at lower rates.

Field-normalized citation rates provide an aggregate measure of men's and women's papers; however, these averages are based on highly skewed distributions. Another lens, therefore, is to look at the proportion of men's and women's papers in certain percentile rankings of citedness and the proportion that are uncited (Table 6.2). For both men and women, articles written individually are least likely to be cited, and those in international collaboration are the most likely to be cited. However, when women are sole or last author, their likelihood to remain uncited is greater than men's. The largest gap, however, appears at the other end of the distribution. There is a 21% difference, for example, in the proportion of men's and women's single-authored works in the top 1% of highly cited papers, with an advantage for men. For men's first-authored international collaborations, that percentage difference increases to 45%. Men's advantages remain within all categories of the 5% and 10% most highly cited articles, though the gap diminishes by the prestige of the percentile. This suggests that a large degree of the citation difference observed may be at the high end of the distribution: that is, that men are more likely to have articles with extremely high citations, thereby influencing the average.

The disadvantages at the top end of the percentile ranking are particularly problematic given the rise of rankings such as Clarivate's "Highly Cited Researchers" list, which is published annually.[47] Several universities now tout this list, using the number of named faculty as an indicator of their success.[48] News agencies contort the meaning of the list, moving quickly from "most cited" to "world's most influential researchers" and noting which institution (Harvard) and country (United States) tops the chart.[49] These news releases do little to provide the caveats of the list or to understand the disparities underlying the indicators. A recent article, however, began to dive into the disparities embedded, noting that, for Asia, few researchers at the top were women.[50] As these rankings gain prominence and begin to be tied to individual and institutional rewards, it is imperative that citation disadvantages are well understood.

TABLE 6.2. Percentage of articles that fall into percentile of citedness and uncitedness of men- and women-led articles, by collaboration level and author position, 2008–2020.

Percentile	Non collaboration		National collaboration				International collaboration			
			First		Last		First		Last	
	W	M	W	M	W	M	W	M	W	M
Top 1% most cited	0.7	0.9	0.7	0.9	0.7	0.9	1.4	2.0	1.4	1.8
Top 5% most cited	3.7	3.9	4.2	4.7	4.1	4.6	6.9	8.5	6.9	8.0
Top 10% most cited	7.5	7.6	8.8	9.6	8.7	9.5	13.6	15.7	13.5	15.1
Uncited papers	34.9	30.8	13.3	13.4	14.5	13.1	10.3	10.4	11.4	10.3

An International Phenomenon

The gap in citations favoring men-led papers is observed in most countries, in terms of both first authorships and last authorships (Table 6.3). Moreover, contrary to what is observed at the world level, we observe a gender gap in citations favoring men for single-authored papers in specific countries, such as Japan (0.53 for women versus 0.61 for men), Germany (0.76 versus 0.98), France (0.54 versus 0.65), and Spain (0.56 versus 0.71). Despite the relative equality of the citation rates for men's and women's single-authored articles at the global level, a citation advantage for men's articles is observed in twenty-eight out of the forty most productive countries.

When analyzing national collaboration, we observe larger differences between the citation impact of men's and women's articles, and those are observed in most countries, and for both first and last authorships. Taking the United States for instance—which has relatively high levels of national (as compared with international) collaboration—we observe a difference of about 10% between the impact of men and women, for both first and last authorships.[51] An important gap in favor of men of about 10% is also observed for Denmark and Switzerland—two countries with high scientific capacity that, contrary to the United States, exhibit relatively low levels of national collaboration.

When considering papers with international collaboration—which generally obtain higher citation rates—the situation is even more pronounced. Men have a citation advantage in first-author positions across all of the forty most productive countries, and there are only three for

TABLE 6.3. Average of relative citations of men- and women-led papers, by collaboration level and author position, by country (for the top forty most productive countries), 2008–2020.

| Country | No collaboration | | National collaboration | | | | International collaboration | | | |
| | | | First | | Last | | First | | Last | |
	W	M	W	M	W	M	W	M	W	M
United States	0.92	0.95	1.16	1.27	1.13	1.24	1.55	1.80	1.47	1.61
China	0.73	0.79	0.96	1.01	0.94	1.03	1.36	1.52	1.36	1.51
United Kingdom	1.08	1.06	1.20	1.24	1.19	1.22	1.53	1.72	1.42	1.53
Japan	0.53	0.61	0.63	0.70	0.67	0.68	0.93	1.07	0.99	1.04
Germany	0.76	0.98	0.95	1.02	0.94	1.01	1.29	1.48	1.26	1.36
Italy	0.70	0.74	0.95	1.01	0.97	0.99	1.22	1.43	1.24	1.34
Canada	0.90	0.93	0.96	1.02	0.97	0.99	1.32	1.50	1.26	1.37
South Korea	0.53	0.55	0.71	0.73	0.72	0.74	0.98	1.08	1.12	1.10
India	0.64	0.65	0.68	0.75	0.67	0.76	1.03	1.14	0.99	1.11
France	0.54	0.65	0.93	0.99	0.92	0.97	1.27	1.42	1.18	1.30
Australia	0.89	0.92	1.03	1.12	1.04	1.09	1.43	1.64	1.43	1.49
Spain	0.56	0.71	0.87	0.88	0.87	0.89	1.17	1.29	1.13	1.25
Brazil	0.39	0.43	0.59	0.60	0.58	0.60	0.88	1.00	0.84	0.97
Russia	0.29	0.37	0.30	0.33	0.28	0.34	0.67	0.77	0.72	0.84
Netherlands	1.24	1.08	1.17	1.24	1.18	1.19	1.44	1.70	1.40	1.50
Turkey	0.48	0.62	0.53	0.56	0.51	0.56	0.88	1.05	0.89	1.14
Iran	0.54	0.77	0.76	0.82	0.72	0.82	1.08	1.31	1.07	1.28
Taiwan	0.61	0.58	0.71	0.71	0.71	0.72	0.91	0.98	1.00	1.02
Poland	0.47	0.51	0.56	0.57	0.57	0.57	0.83	0.96	0.90	0.93
Switzerland	0.82	0.95	1.12	1.27	1.13	1.23	1.45	1.66	1.44	1.55
Sweden	0.86	0.91	0.95	1.03	0.96	1.01	1.24	1.48	1.21	1.33
Belgium	0.77	0.87	1.08	1.09	1.09	1.09	1.32	1.45	1.24	1.33
Denmark	1.08	1.00	1.03	1.17	1.02	1.12	1.31	1.49	1.25	1.44
Israel	0.79	0.72	0.82	0.85	0.81	0.86	1.22	1.24	1.18	1.26
Mexico	0.41	0.43	0.53	0.50	0.52	0.51	0.78	0.80	0.77	0.83
Portugal	0.63	0.63	0.93	0.89	0.94	0.89	1.10	1.15	1.08	1.14
Austria	0.75	0.80	0.92	0.93	0.90	0.93	1.22	1.34	1.14	1.23
Czechia	0.47	0.52	0.60	0.63	0.58	0.65	0.88	0.97	0.83	0.99
Finland	0.96	0.87	0.92	0.93	0.98	0.90	1.18	1.32	1.23	1.19
Norway	0.94	0.92	0.95	0.97	0.95	0.95	1.22	1.35	1.28	1.26

which they do not have a last-author citation advantage (South Korea, Finland, and Norway). This suggests that, despite progress in some countries, citation disparities are pervasive across the globe. These can also be related to the disparities observed in mobility (see Chapter 5).

The Glass Ceiling on Citations

One can always argue that the citation advantage for men is not a bias but merely a reflection of the greater "quality" of works authored by men. To truly test such an assumption, one would need a measure on the inherent quality of a given work. Unfortunately, assessing the true value of a scientific work is no simple task. Another way to approach the problem is to use the Journal Impact Factor (JIF) of the journal in which a paper is accepted as a selection indicator to evaluate the perceived value of the article.[52] Despite its well-known limitations, the JIF is used as an indicator of the reputation and visibility of a journal.[53] This is due, among other reasons, to the relationship between impact factors and acceptance rates, with high-impact-factor journals having significantly lower acceptance rates. Therefore, we can use this as an indicator not of quality, per se, but of selectivity.

Considering the citation disparities depicted earlier, one might expect the gender gap in terms of the JIF and citations to be similar. Yet our results on the differences in men's and women's citations and impact factors show that, for both first- and last-authored papers (Figure 6.2), the average impact factors for men and women are much closer to parity than citations and that, systematically, the gap in citations is much greater than the gap in impact factors, and always in favor of men. Moreover, in some disciplines—such as engineering, earth and space sciences, and biology— women publish in journals with higher impact factors than men yet continue to obtain lower citation rates. It is notable that among these, engineering has some of the lowest rates of authorship for women (less than 20% in most specialties), yet women publish in journals with higher JIFs. Conversely, in the field of health—where women account for the majority of researchers—there is a stronger agreement between the JIFs of journals and the citation of articles. As shown in the insets in Figure 6.2, there is a small relationship between the percentage of women authorship in a discipline and the alignment between the JIF and citation rates, which may suggest that such biases in citations are less likely to occur in fields where

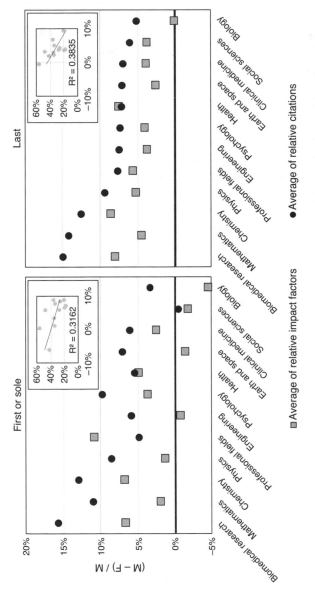

FIGURE 6.2. Differences between men and women in average of relative impact factor and average of the relative citations for articles written as first or sole (left panel) and last (right panel) author, by discipline, 2008–2020. Insets: correlation between the percentage of women in the discipline (y axis) and the relative difference between normalized impact factor and average of the relative citations of men and women.

153

gender parity has been reached. This reinforces the findings (from more than three decades ago) of economist Marianne Ferber, who found that citation gaps tend to diminish as the proportion of women in a field increases.[54]

One may argue that averages are not appropriate measurements, given the skewness of citation indicators. Therefore, to provide a more nuanced analysis of the situation, we grouped articles by women and men according to deciles of field-normalized JIFs and compared gendered citation rates of articles within each decile. The results demonstrate that the largest citation disparity occurs in those journals with the highest impact factors (Figure 6.3). This shows that, even when women publish their works in the most selective journals, their work is less cited than men's, in both absolute and relative terms. This suggests that citations are not distributed in a completely neutral or meritocratic manner and that the disparity in citations is unlikely to be explained by quality differences.

Publication data demonstrate outcomes after acceptance. There may be, however, other selection effects on who submits to these elite journals. We therefore conducted a survey of prolific authors who published between 2008 and 2017 in journals indexed in the Web of Science. Data from more than 6,000 respondents provided insight into the relationship between gender and manuscript submission, rejection, and acceptance rates in *Science, Nature,* and *Proceedings of the National Academy of Sciences.* We found that men were more likely than women to submit to these elite journals and, among those who submitted, were more likely to submit more frequently. Using self-reported data, however, there was no difference in rejection rates between men and women. When women were asked why, therefore, they did not submit to these elite journals, the responses were striking examples of self-perception. Women were more likely to indicate that they did not submit their work because it "was not ground-breaking or sufficiently novel," whereas men were more likely to state that the "work would fit better in a more specialized journal." Women were also likely to indicate that they were advised against submission to elite journals. This may suggest a difference in perceived quality of research; however, results on this were mixed, with women in the natural sciences and engineering ranking their research as being of lower quality compared with their peers, whereas this was not the case in the medical or social sciences. Taken together, these results indicate that women are highly selective in the articles they submit, which suggests both self-imposed and socially amplified disadvantages: women are less likely

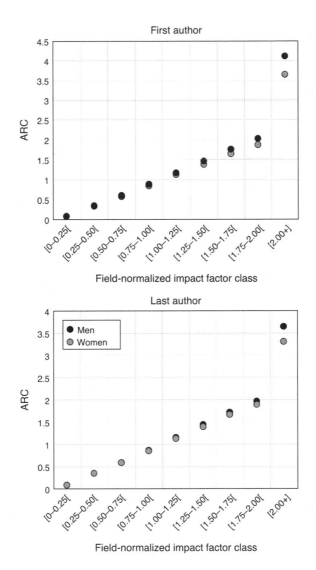

FIGURE 6.3. Average of relative citations (ARC) of men- and women-led articles, according to the average of relative impact factor class, for articles written as first (top panel) and last (bottom panel) author, 2008–2020.

to submit articles for publication and, when they are accepted, their work is not cited at similar rates to men's contributions.[55]

The Role of Self-Citations

It has been argued that self-citations strongly contribute to the gendered differences in citation: that is, that men's higher rate of citedness is due to their increased likelihood to cite their own work.[56] Studies, however, are inconclusive.[57] Using author-level self-citations—that is, the percentage of all citations received by a researcher that they have made themselves—provides insights on the share of citations that authors receive from their own work (Figure 6.4). Specifically, the y axis presents the percentage of self-citations of researchers, as a function of the total number of citations received by researchers (x axis). As shown, men are more likely to have a higher self-citation percentage, irrespective of their total number of citations. After 500 total citations, however, the percentage of self-citations stabilizes at about 6% for women and 8% for men. This finding confirms that men self-cite to a higher degree; however, the difference is not enough to explain the much larger differences in citedness observed earlier in this

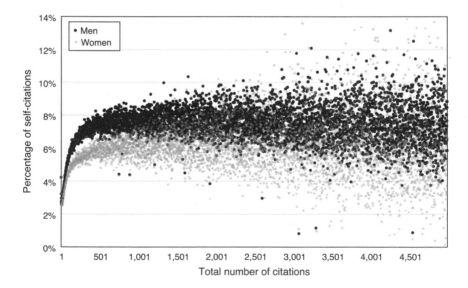

FIGURE 6.4. Percentage of self-citations (compiled at the individual researcher level) of men's and women's papers, as a function of their total number of citations, 1980–2018.

chapter. Therefore, there is a larger mechanism explaining the difference in citedness beyond self-citation.

As noted in Chapter 2, men are more likely to construct teams with other men. This gender homophily in scientific teams has also been reinforced in several other studies.[58] It stands to reason, therefore, that there might be an inherent compound disadvantage: if men cite themselves more, then they may also benefit from the self-citations of other men within the teams they compose. This is precisely the question that physicist Dani Bassett and their team investigated.[59] They found that a relationship exists between gendered coauthorship networks and citation behavior; approximately two-thirds of the observed overcitation of men by other men remains after accounting for the structure of authors' coauthorship networks. Therefore, while some disparity can be accounted for by the disproportionate degrees of self-citation among men and their same-gender coauthors, these mechanisms also do not fully account for the observed differences.

Citing Matilda

Science is not a meritocracy. There are many factors that affect who gets to do science and who receives credit for it. Some of them are institutionalized. The Nobel Prize, for example, is capped at three individuals per prize, although most scientific teams far exceed this number. Given the disparities in production, collaboration, and contributorship, it is no surprise that women are often omitted from this triad of prize winners. Credit, especially in the form of citations, is reinforcing. The concept of the Matthew effect in science was introduced by American sociologist Robert K. Merton, who recalled the biblical verse in the book of Matthew, "For whomsoever hath, to him shall be given, and he shall have more abundance; but whomsoever hath not, from him shall be taken away even that he hath" (Matthew 13:12).[60] Merton focused on how those with preexisting symbolic capital in the scientific community are often the ones to receive disproportionate credit for new discoveries or successes, to the detriment of their collaborators. The same phenomenon has been called cumulative advantage and preferential attachment.[61] Simply, it is the observation that the rewards of science tend to concentrate around certain individuals, labs, and countries and that once success is established, it becomes easier to generate future successes.

However, the second half of the biblical verse is often forgotten. It is not only the accumulation of advantage but also preferential disadvantages for those who "hath not." (The obvious irony is that the literature has focused primarily on the success formulation in this theory.) Historian Margaret W. Rossiter refers to the perpetuation of disadvantages as the Matilda effect in science.[62] As she writes, "If science is to be meritocratic and the history of science to reflect this, similar or equal achievements should receive similar reputations or recognition."[63] Rossiter named the Matilda effect after a nineteenth-century American feminist, suffragist, and early sociologist of science, Matilda Joslyn Gage (1826–1898). Among her many notable works, Gage authored an article titled "Woman as an Inventor," published in 1883 in the *North American Review.* In this work, she reviews the contributions of women inventors across time, noting that "ancient tradition accords to woman the invention of those arts most necessary to comfort, most conducive to wealth, most promotive of civilization."[64] Economic consequences receive a fair share of attention in the article. For example, Gage describes the introduction of the straw bonnet by Betsy Metcalf, the contributions of Catharine Littlefield Greene to the design of the cotton gin, and the perfection of the mower and reaper by Ann Harned Manning. Each of these innovations translated into tremendous wealth in the United States. Gage contrasts this wealth with disparities in intellectual property rights: "In not a single State of the Union is a married woman held to possess a right to her earnings within the family; and in not one-half of them has she a right to their control in business entered upon outside of the household. Should such a woman be successful in obtaining a patent, what then? Would she be free to do as she pleased with it? Not at all. She would hold no right, title, or power over this work of her own brain."[65]

Gage's work speaks to the lack of capital awarded to women and the ways in which economic and symbolic capital is often transferred to dominant social parties (such as Eli Whitney with the cotton gin). It is therefore both tragic and appropriate that Gage was eponymized by Rossiter to represent the underrecognition of women, particularly in knowledge-intensive industries.

Women remain underrecognized for their contributions more than a century after Gage. Today, there remain disparities in the reception of women's work, particularly in some of the most cited scientific journals. Even when subject to the same process of peer review—lauded as the gold standard for evaluating the quality of a researcher's paper—the work of women is cited at a lower rate. These patterns are observed globally, demonstrating the pervasive nature of gender disparities in scholarly impact. It may be argued that these disparities are the result of differences in topics or other properties of research papers. However, studies have shown that, when controlling for all the properties of a paper, women's work remains grossly undercited.[66] We demonstrate that self-citations play only a small role in explaining this difference. There is other evidence to suggest that men's citing behaviors—not only to themselves but to other men as well—are also an explanatory factor.[67] However, this too does not account for most of the disparity. Mounting evidence reinforces the relative and absolute neglect of women's work, despite progress toward equity in science. This has strong implications for scientific careers given that citations are used in the evaluation of scientists for career progression and the allocation of resources and rewards.

Gage's name has been employed again recently in the construction of the Gage Database, which markets itself as the "world's largest directory of women and gender diverse folks in science, technology, engineering, math and medicine."[68] Founded by conservation scientist Katarzyna Nowak, neuroscientist Liz McCullagh, and climate change scientist Jane Zelikova, the platform seeks to expand the visibility of women scientists by providing an accessible list for journalists and conference organizers. In this way, the platform seeks to counter the Matilda effect by increasing the number of women in highly prestigious positions in science.

Another approach to address citation imbalances is "citational justice"—that is, an intentional practice of diversifying reference lists. For example, the Cite Black Women campaign, led by anthropologist Christen A. Smith, encouraged academics to "critically reflect on their everyday practices of citation and start to consciously question how they can incorporate Black women into the CORE of their work."[69] Several other fields, from education to human-computer interaction, have also published calls for greater reflexivity in how researchers are citing.[70] One suggested strategy has been to include citation diversity statements with published research.[71] Proponents argue that reflexivity is an initial step toward reparative justice in citing.[72] Some may opine that these approaches violate the value of meritocracy in science. However, this may actually reaffirm meritocracy through more responsible and intentional practices. Researchers tend to cite the articles that are most visible rather than most relevant. Encouraging scientists to move beyond their expected venues and known institutions to a broader array of literature would be both more equitable and scientifically robust, diversifying visibility rather than privileging certain populations.[73]

We may also want to rethink our entire approach to the utilization of citation as a metric of prestige. Marie Mattingly Meloney, the American journalist who championed Marie Curie, once asked her to nominate Lillian Gilbreth (see Chapter 2) for a Nobel Prize. Curie responded that she would be obliged, if only she knew for which Nobel to nominate Gilbreth, as she fell outside the Nobel categories.[74] This is perhaps part of the problem: we do not know how to categorize the achievements of women. They fall outside our notions of excellence—categories set up in the century of men of science, when single authorship and monodisciplinarity were the modus operandi. Scientific practices, however, have altered dramatically. Instead of continuing to find ways to amend the current system, we should seek a radical reimagining of our system of success. The following chapter, therefore, examines the broad range of contextual factors that mediate women's success in science, which are imperative to consider as we move toward a more inclusive scientific ecosystem.

Chapter 7

Social Institutions

The majority of women scholars in the early twentieth century taught at women's colleges, where they were strictly forbidden from marrying.[1] Professorships—and professional research careers more generally—were not seen as compatible with women's domestic obligations. For example, Harriet Brooks, a graduate in physics tutoring at Barnard College—now part of Columbia University—notified Dean Laura Gill that she intended to marry but wished to stay on as a lecturer. Gill refused, citing the trustees' rule on the employment of married women: "The College cannot afford to have women on the staff to whom the college is secondary; the College is not willing to stamp with approval a woman to whom self-elected home duties can be secondary." In Gill's words, she could not expect a married woman to "carry on two full professions at a time."[2]

Ethel Puffer found herself in a similar situation. She graduated from Smith College in 1891 and had completed all requirements for the doctoral degree at Harvard by 1898. Puffer was the second woman to complete the requirements, having been preceded three years earlier by psychologist Mary Whiton Calkins. Harvard, however, refused to officially grant a degree to either woman, on the basis of gender. It was not until 1902, when the Radcliffe Graduate School for women was opened, that both were offered a Radcliffe doctorate. Puffer accepted, but Calkins refused—it would be a Harvard degree or none at all. Unfortunately, Calkins died long before the policy was changed (1963), making Puffer the first woman to be granted a degree from the Harvard/Radcliffe establishment.

Puffer published her dissertation as a book in 1905 (*The Psychology of Beauty*) while working as a laboratory assistant at Harvard and teaching at several women's colleges (including Wellesley, Smith, and Simmons). However, her career met resistance in 1908 when she decided to marry. The president at Smith, upon hearing of her engagement, warned Puffer that the news had altered his ability to recommend her for a position at Barnard.[3] The marriage ended her academic career. Puffer devoted herself to social activism, focusing on the suffrage movement and other gender-related issues. In the 1920s, she wrote for the *Atlantic Monthly* on the incompatibility of marriage and careers. In her words, "Unmarried women, limited in numbers and in contacts with life, cannot charge the citadel of professional privilege in sufficient volume and momentum to carry it. Until all women of ability, in the sense in which it may be said of all men of ability, are in action, it is probable that few women will reach the highest, and the avenues will remain obstructed."[4]

Progress was slow across the century: 18% of women in the 1921 edition of *American Men of Science* were married, but with vast differences by field (half of the women anthropologists were married, but none of the physicists).[5] By 1938, more than a quarter of all women scientist were married. In an article penned in 1970, crystallographer Kathleen Lonsdale noted that half of women researchers in Britain were single, whereas only 10% of the women in the population were unmarried.[6] Rates of childbearing among scientists are even lower. Several studies published over the last two decades show that college-educated women are less likely to have children and more likely to have fewer children than their less educated counterparts.[7] This is particularly true for women employed in academic work.[8]

There are, however, notable examples of mother-academics, of which the Curie women are perhaps the most famous. Irène Joliot-Curie (1897–1956) was the eldest daughter of Marie Curie and emulated her mother both as a physicist and as a parent. In many ways, Irène had an ideal childhood for a burgeoning scientist. When Marie was awarded her second Nobel Prize in 1911, Irène accompanied her on the forty-eight-hour train ride to Stockholm. The family took holidays with Albert Einstein's family. She was schooled in a private cooperative with the children of other leading French academics, rather than in the public school system. When she was in her early twenties, Irène accompanied Marie to the United States, where she delivered lectures, received honorary doctorates on behalf of her mother, and was welcomed to the White House by President Warren G. Harding.[9]

Irène's connection to her mother intensified in 1916, when she left school to aid her in the war efforts. For four years, Irène and her mother took x-ray equipment onto the battlefields to help treat injured soldiers. This work would garner Irène a military award; however, it is also suspected that this repeated exposure to x-rays is what led to her untimely death from leukemia in 1956. Irène continued her work on radium after the war, publishing her first paper on the subject in 1921.[10] Although she did not collaborate with her mother, she worked in the same lab and benefited from the network that her mother had established. When she defended her thesis at the Sorbonne in 1925, more than a thousand people were in attendance and reporters from across the globe covered the event. Her doctoral adviser was Paul Langevin, who had been supervised by her father, Pierre, and had been romantically involved with her mother.

It was Langevin who introduced Marie to Frédéric Joliot, a young physicist with no graduate degree who was looking for a position as a junior assistant. Marie hired Joliot and placed him under the tutelage of Irène, three years his senior. Their working relationship eventually developed into a romantic one and they married in 1926. Irène celebrated her wedding with a luncheon and returned to the lab that afternoon. Her approach to childbirth was similar: she worked through the morning and went to the hospital in the afternoon, giving birth to her daughter, Hélène, in 1927. Irène's tuberculosis was confirmed during childbirth, and the doctors ordered rest and no further pregnancies. The willful Irène returned to work immediately, attending the famous Fifth Solvay Conference in Brussels just weeks after giving birth. She had her second child, Pierre, in 1932. Her health declined significantly after giving birth, and she would spend extended times away at retreats, not unlike her mother. Despite this, Irène is noted as saying about motherhood, "I would have never forgiven myself for having missed such an astonishing experience."[11] In many ways, Irène reflects the contemporary expectations for women scientists. As was articulated by the Royal Society in 1971, "She must be a good organizer and be pretty ruthless in keeping to her schedule, no matter if the heavens fall. She must be able to do with very little sleep, because her working week will be at least twice as long as the average trade-unionist . . . She must be willing to accept additional responsibility even if she feels that she already has more than enough. But, above all, she must learn to concentrate in any available moment, and not require ideal conditions in which to do so."[12] For women to be ideal workers in science, they must do so not at the expense of other social expectations— as wives, mothers, and social actors—but in addition to these obligations, rarely without the infrastructure to support these competing labor roles.

As shown throughout this book, historical and present-day disparities are the consequence of persistent and systemic structural barriers to women's advancement in science. The current situation is a consequence of deeply rooted systems that, we would argue, are the result of intersecting disadvantages: cultural demands on women as caregivers, conceptualization of scientific work as masculine, overt discrimination and harassment, and institutional structures within science. Science is an inherently social activity that is affected by the norms of the societies in which it evolves. Addressing disparities and inequities in science, therefore, requires a complex understanding of the more general context in which scientific work is conducted. Furthermore, the social structures that lead to stereotypes and discrimination are compounded at the intersection of gender and other vulnerable identities. Race, sexual orientation, and disability status, for example, all contribute systemic barriers to participation in science. Policy recommendations that ignore these factors are unlikely to be successful. Before concluding our book, we examine some of the social and institutional structures that serve to perpetuate disparities and the interventions that might be used to mitigate these.

The Incompatibility of Parenting

Most families are not like the Curies. Studies on family structures have repeatedly reinforced that when women have children, they remain the main caregivers, even in dual-earner households.[13] This is particularly true when children are young.[14] As the Bureau of Labor Statistics Time Use Survey reported, "In households with children under the age of six, men spent less than half as much time as women taking physical care of these children."[15] Qualitative studies in the academic realm have confirmed that academic women's work follows other sectors, with disproportionate expectations of care work for working academics.[16] Men are engaging more with domestic responsibilities than in previous generations, but it has been argued that the contribution of men to domestic labor is not changing as quickly as occupational differences.[17] Moreover, having children was shown to have little to no effect on fathers' incomes. For mothers, however, relative earnings and probability of employment declines with children, and this is particularly the case for women with college degrees.[18]

In science, domestic responsibilities have been heralded as an explanation for differences in productivity, collaboration, and citations. However, results on the relationship between parenting and scientific work have been mixed. Some studies have found that women with children face a productivity penalty compared with men with children and women without children.[19] Other studies have suggested no association between production and family obligations or an increase in productivity immediately after birth.[20] The latter may be an artifact of publication lags and the pressures of an academic environment, demonstrating that women accelerate their productivity directly *before* birth, which manifests after.

Sociologist Mary Frank Fox called attention to the complexity of this research area, noting that the type of marriage (first or subsequent), age of children, and other factors must be considered in the analysis.[21] Inconsistencies observed across studies may be the consequence of the wide differences in methods used and populations studied.[22] For example, one oft-cited study examined chemists who graduated between 1955 and 1961. In their analysis, the authors examined all women who were listed as graduates in the American Chemical Society: the full population was only 231 women. This analysis is deeply contextualized in time, geography, and discipline, and while it provides some evidence, generalizing to the present time is highly problematic. Other studies suffer similarly from high degrees of localization and context specificity.[23]

Furthermore, many previous studies relating parenting and productivity take for granted that having a child equates with engagement in parenting labor and "do not distinguish how parenting roles are different both within and outside of gender roles."[24] Women researchers are more likely than men to have partners who are employed full time, to spend more time on domestic activities than counterparts who are men, and to sacrifice research time, rather than teaching or service obligations, in order to do more household activities.[25] Furthermore, to characterize labor, it is essential to examine not only the workweek but also the weekends, particularly for scientists, who tend to have nonstandard work schedules.[26]

The uncertain state of research prompted us to reexamine the relationship between parenting and scientific production.[27] The goal of the project—led by science study scholar Gemma Derrick—was to analyze the relationship between parenting engagement and research production, taking into account country of origin and discipline, and to incorporate scientific impact into the model (an oft-neglected status symbol when examining the effect of parenting on scientific careers).[28] Furthermore, given the changing nature of dual-earner households and the increased participation of men in childcare, we wanted to examine not only women but also men, to more comprehensively understand the relationships between parenting labor and academic labor.

Using authors who had published at least once in a journal indexed by the Web of Science as a sampling frame, we conducted a global survey with responses from more than 10,400 publishing parents. This provided baseline demographic information on how the mean number of children—for those who have at least one—varies by discipline, age of researcher, and country of origin (Figure 7.1). We found stronger variation in men across disciplines, with those in clinical medicine having the highest number of children, and those in earth and space sciences having the lowest. It is notable that these disciplines are the least and most mobile, respectively (Chapter 5), reinforcing the relationship between mobility and parenting duties. It is also worth noting that the disciplines with the highest numbers of children align with Pierre Bourdieu's analysis from the 1960s of number of children by occupation. Drawing on data from a survey by INSEE (the French National Institute of Statistics and Economic Studies), Bourdieu provided statistics on the number of children by occupation of the father in the 1960s.[29] He showed that professions associated with middle-class status are likely to have fewer children than those in the lower and higher classes, as a way to maximize the upward mobility of their offspring. More specifically, those in liberal professions—such medical doctors and lawyers—had the highest number of children, along with those without any high school diploma.[30]

Our survey also demonstrated differences by country of affiliation, with, for example, Danish men having more children than men from Portugal. Greece and Portugal were the only two countries where women scholars had more children than men. Polish women had the lowest number of children, on average, which is striking given that this is one of the countries with the greatest representation of women among authors (Chapter 1). The gap between men's and women's number of children has decreased over time: men in the older cohorts had more children than women who began their careers at the same time, but this gap is virtually nonexistent for the current cohort of scholars. That is, contemporary men and women scholars are having relatively similar numbers of children.

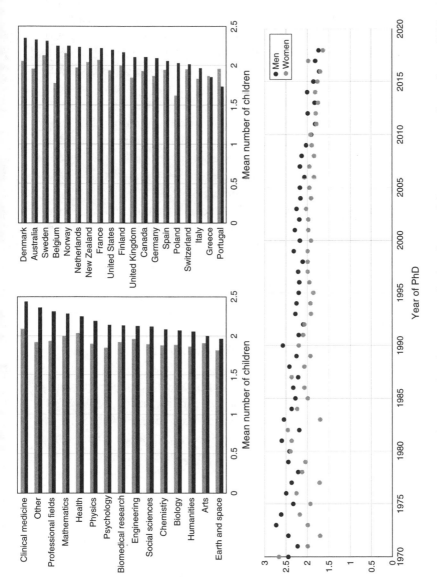

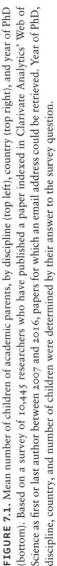

FIGURE 7.1. Mean number of children of academic parents, by discipline (top left), country (top right), and year of PhD (bottom). Based on a survey of 10,445 researchers who have published a paper indexed in Clarivate Analytics' Web of Science as first or last author between 2007 and 2016, papers for which an email address could be retrieved. Year of PhD, discipline, country, and number of children were determined by their answer to the survey question.

However, despite having the same number of children, there is still a gap in parental *engagement.*

Our survey demonstrated that the research production penalty for men and women is a function of the level of engagement in parenting activities, measured as the times at which one is active in parenting and the types of tasks that are done. Overall, our analysis revealed that parenting comes at a cost to the production of research articles: the more engaged a scientist was in caregiving, the lower their production. This cost, however, is largely borne by women who are both more likely to be primary caregivers and to be more highly engaged even in self-proclaimed shared or "satellite" roles—that is, to have an equal or secondary role in parenting activities. These results suggest that parental engagement is a more powerful variable to explain gender differences in academic productivity than the mere existence of children, with strong policy ramifications.[31]

The survey also provided insight into academic couples, which constituted nearly a third of the respondents.[32] As noted in Chapter 5, women are more likely to have an academic for a partner than men.[33] These academic couples arguably experience similar pressures and professional labor expectations. However, when asked for levels of agreement with the statement, "My partner/other takes on the majority of childcare so I can focus on my career," men were more likely to agree and women to disagree. Similarly, a higher proportion of men than women strongly disagreed with the statement, "I take on the majority of childcare so that my partner/other can focus on their career." Men's family behaviors have not kept pace with the advancement in women's changing occupational demands; this, combined with the inflexibility and masculine normativity of workplace environments, has created severe constraints on women. Simply put, women still bear the burden of domestic care while navigating the same expectations as men in their workplaces. This is particularly exemplified in the case of academic couples claiming shared parenting, wherein one would expect equality in both academic and domestic expectations.

Parental leave has been the primary way in which institutions have provided support to counterbalance parenting demands.[34] Another mechanism employed by some institutions and funding agencies is a so-called extra hands award. This funding allows women to hire additional support—such as technicians and postdoctoral fellows—when they become primary caregivers.[35] This is essential to ensure that women are allowed time for childbirth and recovery, but it also acknowledges the disproportionate burden they carry in terms of childcare. Furthermore, extra hands awards alter the nature of support from a model that removes women from the lab (literally "leave" initiatives) to models that allow them to stay active in their research while parenting.

Immunologist Laurie H. Glimcher advocated for such assistance at the National Institute of Allergy and Infectious Disease during her presidency of the American Association for Immunologists.[36] HANDS-ON: Primary Caregiver Technical Assistance Supplements was launched in 2004 to provide caregivers funding supplements to hire technicians. The program remains active.[37] A similar program has been operating since 1997 at Massachusetts General Hospital: the Claflin Distinguished Scholar Awards provides bridge funding for junior scholars who serve as primary caregivers.[38] A retrospective evaluation from the first ten years suggests that it served as a positive intervention, helping to advance women into senior positions in biomedicine.[39] The success of the program led to adoption at other institutions, such as the University of Pittsburgh, the University of Massachusetts, and Stanford University.[40]

Organizations have also begun to provide funding for assistance not only in the lab but in the home. For example, the Christiane Nüsslein-Volhard Foundation in Germany provides funding for household chores and childcare, stating that the "time thus freed allows [women scholars] to continue working at a high standard, despite the double burden."[41] Funders are now following suit: the National Institutes of Health provide funding for doctoral students and postdoctoral fellows with children to offset the costs of childcare.[42] Several institutions provided short-term assistance for childcare during the COVID-19 pandemic, recognizing the cost to women's scholarship as children returned to the home during lockdowns.[43] To achieve true equity, institutions should recognize that this burden, while amplified during the pandemic, persists beyond it, and provide adequate support to allow women to remain active in research while parenting.

Masculine Stereotypes of Science

Social institutions both reinforce women's roles as caregivers and emphasize the masculinity of scientific work. Stereotypical representations of scientists in the media evoke familiar images: lab coats, chemistry beakers, telescopes, and a man behind it all. This representation has not changed much since anthropologists Margaret Mead and Rhoda Métraux surveyed high school students in the 1950s: "The scientist is a man who wears a white coat and works in a laboratory. He is elderly or middle aged and wears glasses."[44] The stereotypical and sexist image is perpetuated in public representations as well as within the most prestigious scientific magazines. For example, a study of *Science* and *Nature* demonstrated that women are underrepresented in visual depictions in advertising, featured scientists, and stock photographs.[45] Similar results demonstrating the prevalence of male casting in the portrayal of scientists were found in a comprehensive examination of web content, from BBC to blogs. When women are given a voice or representation, they are often pictured in communal roles of collaboration and teamwork, rather than amplifying individualistic characteristics of drive or brilliance.[46]

Gendered representations in media not only privilege men but reinforce implicit and explicit biases against women. For example, one experiment asked students in an undergraduate biology classroom to judge the competence of their peers. In the first week, there was no explicit gender difference in the response. By week seven, however, men were ranked as having a higher mastery of content than women, even when controlling for class performance and outspokenness.[47] A similar experiment was done with scientific abstracts, in which reviewers were intentionally given abstracts with false men and women authorships. The scholars found that abstracts by "men" were rated as having significantly greater scientific quality.[48]

The same is true online. An analysis of the contributions of 1.5 million computer programmers publishing on the GitHub platform showed that, when the first name of authors provided an indication of gender, codes produced by women were likely to be rejected more by their peers, whereas the opposite was observed when only the initials of the first name of the authors were presented.[49] This suggests that, a century after Lillian Gilbreth had to mask her gender using initials (Chapter 2), the use of given names by women can still lead to bias and inequities. The use of initials is overrepresented in fields that are seen as less "suitable" to women, due to notions of inherent talents necessary to be successful.[50] Women are also judged differently on the basis of their profile pictures alone. In a study of three academic social media platforms (Mendeley, Microsoft Academic Search, and Google Scholar), we found that, when controlling for all aspects of the photo (for example, the amount of visible skin, the presence or absence of glasses, and the color of clothing), men were more likely to be perceived as "professional" and women as "attractive."[51]

Several initiatives have focused on changing the stereotypes about the typical scientist. The #iamascientist project, for example, focused on creating resources for K–12 classrooms that celebrate diversity in STEM.[52] Resources provided include story kits of successful scientists, posters that highlight diverse representation, and lesson plans for teachers. The material reaches more than 350,000 students each year across all fifty US states. Hashtags have also been used to amplify the work of women on social media, such as #womeninscience, #womeninSTEM, and #ILookLikeAnEngineer. This form of "hashtag activism" creates a visible stream of photos and activities amplifying diverse representations in scientific careers. The utilization of these hashtags has been derogatively termed "slacktivism," but these simple acts can serve as powerful vehicles for changing conversations and creating communities.

Linguist Reem Alkhammash has shown that the use of hashtags like #womeninSTEM "contributed to increasing the strength of women in the STEM community in social media, evidenced by their practices of advocacy, networking, and challenging gender biases online." She also argued that the use of hashtags is a form of "discursive activism that focuses on the larger dialogue of women in STEM and highlights dominant forms of sexism and gendered stereotypes of women's work in male dominated professions." This work, Alkhammash argues, serves to "raise awareness, build allegiances, resist stereotypes and offer solidarity for women."[53] In addition, some hashtags have explicitly sought to engage allies in this work—for example, the #HeForShe hashtag, which came from the program of the same name developed by the United Nations.[54] The same is true for hashtags for other minoritized communities, particularly when they co-occur with other well-networked hashtags. For example, in a study of transgender voices on Twitter, communications scholar Brooke Foucault Welles and colleagues demonstrated how the hashtag #GirlsLikeUs played "an indispensable role in constructing an intersectional networked trans community while simultaneously shifting representations of trans people that arise in the public sphere."[55]

Wikipedia also serves as a reference point for amplifying the work of women scientists. Nearly ten years ago, Emily Temple-Wood, then an undergraduate student at Loyola University, began to write biographies of women in science on Wikipedia.[56] In 2014, she founded the WikiProject Women Scientists, which is a group dedicated to providing high-quality coverage of women scientists on Wikipedia. Between 2014 and 2018 the proportion of women across all Wikipedia biographies increased from 15% to 17.7% of the total. A similar enterprise was initiated by British physicist Jess Wade, who, as of February 2019, had written 500 biographies of women scientists on Wikipedia.[57] These statistics demonstrate progress, but work remains to be done. Wikipedia still represents fewer women than traditional encyclopedias: arguably, given the contemporary nature of Wikipedia, this should be reversed. One explanation is that Wikipedia holds women to a higher standard of "notability"—the requirement that kept Donna Strickland from being included just months before her receipt of the Nobel Prize.[58] Furthermore, even when women have pages, they tend to be less robust and less frequently edited.[59]

The internet was dominated by men for much of its early years, though the gap in usage has nearly disappeared.[60] Despite these advances, the web is often an inhospitable climate for women, who are significantly more likely to experience online harassment, much of it of a sexual nature.[61] Take, for example, classicist Mary Beard, born in 1955 in the United Kingdom. She lectured at King's College, Cambridge, and Berkeley and was made a dame commander of the Order of the British Empire for her study of classical civilizations. This classical biography may seem unusual for one of the most active scholars on Twitter, with more than 316,000 followers as of this writing. Her engagement on Twitter is merely an extension of her lifelong commitment to public engagement. She blogs, contributes pieces of journalism, and has done several television and radio shows with the BBC. Beard's media appearances have given her great acclaim but have also come at a personal cost. She recounted the extreme forms of cyberbullying she has received on Twitter, from the pornographic, sexist, and misogynistic to the downright violent. At one point, she reported a bomb threat leveled against her on social media.[62]

It has been argued that a "social media presence is just as important as taking the podium at a conference."[63] In 2014, geneticist Neil Hall pushed back on this with the satirical proposal of a "Kardashian index"—a measure of the discrepancy between the visibility on social media and bibliometric indicators of a given scientist. He argued that, although "social media is a valuable tool for outreach and the sharing of ideas, there is a danger that this form of communication is gaining too high a value and that we are losing sight of key metrics of scientific value."[64] His analysis found that most "Kardashians"—researchers with a higher number of followers than citations—were men. However, scholars like anthropologist

Kate Clancy noted that the Kardashian index may have a negative effect, disparaging many who engage heavily in science communication.[65] Despite these challenges, several women continue to engage online: an estimate of women scientists on Twitter suggests higher representation than in the scientific workforce generally, or in the production of scientific articles.[66] The shift in the composition of the online space and amplification of women's voices may serve to create more hospitable and equitable spaces for women's scientific discourse, both online and offline.

Science as a Self-Organizing System

Science is a social institution with its own set of norms; as such, it also bears responsibility for the reproduction of inequities. One of the ways in which the scientific community regulates itself is through peer review, which is performed at many points of the system—from hiring to grants and journal articles. Peer review serves as a quality control for scientific claims or expertise before it is disseminated. To operate properly, peer review must be free from bias. There is ample evidence, however, of systemic bias against women, long preceding the contemporary peer-review practices. Take, for example, Agnes Clerke, who, in 1890, was elected as one of the first members of the British Astronomical Association, the same year in which she published her much-acclaimed *The System of the Stars*.[67] She was also elected a member of the Astronomical and Physical Society of Canada, the Liverpool Astronomical Society, and the Astronomical Society of the Pacific. By 1892, the question was again raised of allowing women into the Royal Astronomical Society (the proposal having been rejected in 1886). The ballot failed, but Clerke was invited to attend meetings (albeit not as a full member). The decision to include women as full members would not be made until 1915, eight years after Clerke's death. She would nonetheless be awarded the Actonian Prize in 1893 by the Royal Institution—a prize for scientific essays that would also be awarded to Marie Curie fourteen years later.

Despite these accolades and recognition, there remained many who believed that women were not well suited for scholarship. The reviews of Clerke's third and last major work, *Problems in Astrophysics* (1903), were generally strong: "So happy, so strong, so useful a book . . . I do not believe there is a man living who knew beforehand all the facts that you have brought together and brought together so well in their proper places."[68] William Wallace Campbell, American astronomer at Lick Observatory, stated, "The appearance of Miss Clerke's book could scarcely have been better timed. By way of suggestions for future lines of research this book is the richest one known to me." The review in *Nature,* however, would not be as kind: "A cynic has said that it is a characteristic of women to make rash assertions, and in the absence of contradiction to accept them as true. Miss Clerke is apparently not free from this weakness of her sex." *Nature* editor Richard Gregory chastised Clerke, stating, "Passengers are respectfully requested not to speak to the man at the wheel."[69]

In 1906, Gregory reviewed a revised version of *The System of Stars.* His opinion of women scientists remained clear: "The intuitive instinct of a woman is a safer guide to follow than her reasoning faculties; and although in these days it is considered ungracious to make this suggestion, evidence of its truth is not difficult to discover in most literary products of the feminine mind. It is no disparagement to Miss Clerke to say that even she shares this characteristic of her sex, so that sometimes she lets her sympathies limit her range of vision in the field of stellar research."[70]

Peer review continues to be problematic for women and other under-represented groups.[71] As sociologists Daryl Chubin and Edward Hackett have noted, "Analysts of peer review at times are blinkered by the powerful values that make up the cultural context of science, hence are unable or unwilling to notice that science operates imperfectly."[72] Several studies have challenged the meritocracy of peer review, demonstrating sustained bias in peer review in which women's work is rejected at higher rates than men's submissions.[73] One explanation for this bias is the combination of lower rates of women as reviewers and the effect of gender homophily in reviewing: that is, both men and women exhibit preferential treatment for same-gender authors.[74] Given that men account for the largest proportion of reviewers, this homophily bias disproportionately favors men. In a comprehensive study of *eLife*—a large open-access biomedical journal—we found that women were far less likely to have their submissions accepted for publication, particularly when they were reviewed by all men.[75] Homophily patterns likewise affect the selection of reviewers: men editors disproportionately invite men reviewers. There are also gendered patterns in recommendation, with manuscripts handled by women editors rejected at higher rates.[76] Other evidence reinforces that women are more critical reviewers, which may not lead to the same types of homophilic advantages for women.[77] Therefore, in addition to providing bias training for reviewers and editors, journals should move toward increased transparency, providing data on the gender composition of authors, reviewers, editorial boards, and editors and continually monitoring and reporting on improvements.

Movements toward transparency have challenged the dominant single-blind review practice, which has been shown to have a negative effect for women.[78] This resonates with some of the earlier findings on women's higher levels of negative reviews: women tend to rate single-blind articles less favorably, whereas men are harsher reviewers for double-blind articles.[79] This may partially explain other studies that have shown that blinding reviews serve to mitigate gender disparities in outcomes.[80] Open peer review has also been suggested as a mechanism for deterring bias, though there is not sufficient evidence to reinforce this. Theoretically, opening the content of reviews to the public should reduce ad hominem attacks and unfair reviews. However, naming peer reviewers could have a chilling effect, reducing the ability of junior and midcareer scholars—who are more likely to be women—to openly critique those in more powerful positions within the field.[81] It is essential, therefore, that journals engage in more experimentation with peer-review interventions, to gain a greater understanding of the mechanisms contributing to disparities in outcomes.

Peer review extends beyond the evaluation of manuscripts for publication. The selection of candidates for positions, awards, and funding is also a space where gender disparities are well established, and these processes have crucial consequences for researchers' careers. For example, hiring committees often make assumptions about the mobility of women relative to men, which harms women in selection for academic positions.[82] Studies presenting senior investigators with identical résumés for candidates with clearly gendered names have shown that both men and women rate women candidates lower and offer them lower salaries.[83] This type of overt discrimination of women candidates was reinforced in a quasi-experimental design of funding that found that gender disparities in funding were explained by lower rankings of the principal investigator, rather than the quality of the proposed research.[84] Taken together, these results suggest that committees should focus on the evaluation of projects rather than the people and their associated accolades. Evaluation of scientists over science will continue to favor majority populations and reproduce cumulative advantages.

Salary Inequities

Disadvantages are faced in both symbolic and economic capital. Consider again the case of Agnes Clerke. Determined to conduct work at an observatory, she applied for and was offered an appointment at the Royal Observatory as a "supernumerary computership," which would have made her the first woman on staff at the Royal Observatory. The duties, however, were listed as "literary" and she worried that she would not be able to make use of the Lassel reflector, which was her main goal. Furthermore, the living conditions were not ideal, as Greenwich Park was considered unsafe for women, particularly at night. The computerships were eventually offered to other women, at a much lower rate than she was offered. Her correspondence indicates her concern about the positions: "I should think no educated woman would accept such a post at such a minute salary (three pounds a month for the inferiors) unless with a view to training for something higher, and getting into a mechanical routine would be fatal to that end."[85] The other computers were young men who would perform the task for a few years before moving on to other positions. In contrast, the four women computers employed from 1890 to 1895 were all products of Cambridge women's colleges who had taken the Mathematical Tripos.[86] The situation has not improved. In the United Kingdom, for example, the Equal Pay Act was passed in 1970. However, a gender pay gap as large as 24% in the scientific and technical activities sector persists today.[87] This wage penalty is particularly felt for women who are also mothers.[88]

Devaluation can also be seen within disciplines. All disciplines that have majority women authorships are rooted in feminized professions—such as nursing, social work, library science, and education (Chapter 1). It stands to reason, therefore, that the associated disciplines would also be feminized. Feminization followed several paths, however, due to the trailblazers in the discipline and the interaction with other disciplines across time. Library science provides a good example of this. Whereas nursing had Florence Nightingale—who was heavily involved with the women's movement—librarianship was largely professionalized by Melvil Dewey, a man with a well-documented history of sexual harassment, racism, and anti-Semitism.[89] Dewey founded the first School of Library Economy at Columbia College in 1887 and enrolled women from the beginning, despite the consternation of the trustees.[90] Women were also featured among the first doctoral graduates from the Graduate Library School at the University of Chicago (established in 1926).[91] According to the MPACT database, Eleanor Stuart Upton received the first library science doctoral degree in the United States in 1930.[92] Despite this, women remained underrepresented among the faculty of library science programs for decades. Forty years after the first doctoral degree was awarded, women were only about 40% of the faculty, a number that was relatively static for the next decade. It was not until 1996 that the number reached parity (49.5% men, 50.5% women), mostly due to the influx of women as assistant professors.[93] The percentage of women among the faculty hit a zenith in 2009 at 55.9% but has fluctuated around parity for the last decade. Heads of school were mostly men until recent decades, not unlike the leadership in libraries.[94] The increasing feminization in leadership, however, has not corrected salary inequities. Women are paid less than men at every level (from lecturers to deans). In 2020, salary equity was closest at the assistant and associate levels, where women made 4% and 3% less, respectively, than men in the same ranks; the largest discrepancy was at the dean and director level, where women made nearly 23% less than men in these leadership positions.

Feminization of the library science professoriate slowly reversed during the last two decades, arguably due to the influx of computer science graduates. In 1997, 8% of professors in traditional library science programs had computer science degrees; by 2015 they represented nearly a quarter of the faculty.[95] The irony of the feminization of library science being threatened by the masculinity of computing is that computing observed the opposite shift across the twentieth century: the earliest computer programmers were women, as were their supervisors.[96] However, despite the success of a few of these early "computers," such as the notable Grace Hopper, much of the labor of programming was marginalized due to the gender and education level of early programmers.[97] By the 1950s, programming had shifted toward a masculine orientation. Historian Nathan Ensmenger has argued that this was, in part, due to the adoption of aptitude tests and professional requirements that discriminated against women.[98] By the 1960s, when computer science began to establish itself as a theoretical discipline in academic chambers, women had been marginalized to the ranks of clerical workers. In 1970, women were less than 15% of undergraduate computer science students; the numbers rose until the 1980s and then declined to around 15% again in recent years.[99] The composition of the faculty reflects these trends: in 2019, women represented less than 16% of full professors and less than a quarter of all tenure-track faculty.[100] This is marked improvement, however, from 2002, when women were at half these rates.[101] Despite both progress and the high market value of these faculty positions, women earn less, on average, than the men in their department at each rank, as is the case in library science.

Salary inequity is visible both within and across disciplines. Differences across disciplines are usually dismissed with the "market value" argument: that is, that the professions that are available to these faculty outside academe—and especially in the business and industrial worlds—pay different rates. Therefore, given that librarians make less than, for example, computer engineers, faculty in these areas should also be paid different amounts. The market value argument, however, replicates the salary inequities of feminized professions within the disciplinary space, which duplicates the salary penalty for women. Not only does this reinforce the devaluation of women's work; it also negates the fact that academic labor expectations do not drastically vary across disciplines in academe, allowing cultural devaluation of professions in the public sphere to influence equity in higher education.

Reimagining the Ideal Worker

Kathleen Lonsdale (née Yardley) (1903–1971) had a remarkable scientific career. By the age of sixteen, she had received a scholarship to London's Bedford College for Women. She received the highest score on the Honours Physics BSc examination, and one of the examiners—Nobel laureate William H. Bragg—invited her to join his lab at the University College London (and, subsequently, at the Royal Institution). In 1936, Lonsdale was awarded her doctoral degree from University College London, and in 1945, she and biologist Marjorie Stephenson were the first women elected as fellows to the Royal Society of London (founded in 1660). In 1949, she was appointed professor of chemistry and head of the Department of Crystallography at University College London (making her the first woman tenured professor at this institution).[102] She became dame of the British Empire in 1956, was awarded the Royal Society's Davy Medal in 1957 (the second woman only to Marie Curie, who received it in 1903), was the first woman president of the International Union of Crystallography in 1966, and became the first woman president of the British Association for the Advancement of Science in 1967. She was also a mother.

Several important interventions made it possible for her to balance domestic and professional responsibilities. At twenty-four, she married Thomas Lonsdale, a textile chemist, and they moved to Leeds for Thomas's job at the Silk Research Association.[103] Thomas was unusual in many ways in postwar Britain, where traditional gender roles had been reasserted. He is noted as saying "he had not married to get a free housekeeper" and encouraged Kathleen to continue her scientific work.[104] The year after they arrived in Leeds, Kathleen, who took a position at the University of Leeds, published what became her most well-recognized piece of research in *Nature;* the full report of the work was published in the *Proceedings of the Royal Society of London* in 1929, the same year she gave birth to her first child. Two more children followed, in 1931 and 1934.

Rather than following in the footsteps of Irène Joliot-Curie, Kathleen chose to stay home with her children during their early years. She did not, however, abandon scientific work. Rather, she shifted her work to something that could be done away from the laboratory, developing reference tables to help scientists more easily derive chemical structures from x-ray photos. She also received help from mentors along the way. For example, Lonsdale's initial mentor, Bragg, continued to support her throughout her career. At least twice during her career he secured small grants for her, expressly intended to pay for a housekeeper so that she could continue her scientific work. Bragg recognized that the greatest way to help retain a talented woman scientist was to provide her resources to reduce her domestic obligations. Lonsdale herself gave this advice to women scientists: "For a married woman with children to become a first-class scientist, she must first of all choose or have chosen the right husband." She defines the right husband as one who would "share in the household chores, accept his wife's hectic schedule, and otherwise do what he could to ease the domestic burdens on his spouse."[105]

Science policy cannot be based on opportunistic domestic partnerships. Therefore, Lonsdale's story reinforces the need for institutions to invest in work-family policies.[106] There have been several advancements in this realm, including parental leaves, stopping the tenure clock, spousal hiring programs, and on-campus childcare centers.[107] However, some well-meaning policies may actually serve to exacerbate inequities. For example, institutions sprang into action around COVID-19, responding to the pressures on women academics during this time.[108] Many of them moved toward tenure-clock extensions. While clock extensions—for COVID or caregiving—can provide necessary space for women, they also have several negative consequences: increasing the time of vulnerability, exacerbating pay gaps, and increasing the bar for productivity if reviewers are not well trained. For example, if institutional requirements annualize productivity, instead of evaluating the portfolio of work, women may face a penalty for prolonged time at rank. This may, for example, explain why men are more likely to receive tenure than women, even when accounting for productivity differences.[109] Institutions should rethink promotion and tenure to ensure transparency and to remove criteria that systematically discriminate against women.[110]

On the whole, criteria for tenure and promotion—as well as the general allocation of symbolic capital—are still based on a view of academic labor that adheres to "ideal worker norms," which reflect occupational and domestic labor distributions that favor married men whose wives are primarily dedicated to domestic labor, and may explain their moderate success at reducing the gender gap.[111] The rigidity of the pipeline model also presents barriers for reentry for those who deviate from this norm or suspend their progression for any reason.[112] Even "family friendly" policies, such as stopping the clock for tenure, reinforce the idea that moments away from academic labor are deviant. Qualitative research has confirmed that women feel more pressure to conform to certain norms and feel greater stress in balancing work and family.[113] Those models are changing as fewer academics—both men and women—fit these norms. However, there remain many interventions that are required to accelerate change and ensure that parity is met with equity.

Chapter 8

Recommendations and Conclusions

We began our book with a simple motivation and underlying assumption: that changing the composition of the scientific workforce has implications for all sectors of society. As philosopher Perry Zurn observes, "Women and their compatriots have inherited not only questions, but lines of questioning and architectures of curiosity that best serve the status quo."[1] An expansion of the scientific workforce, therefore, leads not only to greater equity among scientists and a wider representation of the population within science but to an expansion and reorganization of science. Drawing on theories of sociologist Pierre Bourdieu, we previously remarked,

> It should be recalled that what counts as "legitimate" or "important" research is still a function of the dominant agents of a scientific field (Bourdieu, 2004). Given that men still occupy, more often than not, the dominant positions and participate actively in the formulation of research policies, and that many women also internalized these "dominant" values, it could happen that even in the current reconfiguration of the tasks assigned to universities, domains that are considered "significant" will remain for a long time those of "hard" and "masculine" science. For example, research on the genome is considered more important than nutrition and dietetics, even though it is scientifically plausible that better eating habits are more likely to lower cancer rates, in the medium term, than personalised genetic manipulations . . . It is therefore likely that true equality in research will only be achieved when strategic positions, which impose categories of

thought and evaluation criteria, are occupied by researchers whose research topics are currently being undervalued.[2]

It has been suggested that motivation for diversity in science can be framed using three main ethical frameworks: virtue ethics suggests that scientists should be endowed with characteristics of humility, equity, and generosity that would lead them toward addressing injustices; a deontological approach may suggest a duty and responsibility to engage in the practice of equity; and a utilitarian approach suggests that an expansion of the scientific workforce leads to greater good for the greatest number.[3] Our book reinforces this approach: when science is reflective of the full population, the benefits of science will also extend to the fullest.

We pursued this line of inquiry with a large-scale empirical work, examining intersectionality in science.[4] Focusing on the US population and using categories from the US Census, we assigned gender, race, and ethnicity to authors of a corpus of nearly 5.5 million scientific articles indexed in the Web of Science. Using these data, we examined the relationship between intersectional identities, topics, and scientific impact. We found homophily between identities and topics: that is, that the topics studied by scientists reflect their lived experiences. Therefore, expanding the scientific workforce will also reflect an expansion of the base of knowledge. We also performed a counterfactual analysis of how research topics would have changed had the author distribution over the last forty years matched the demographic population as represented in the 2010 US Census. This scenario would yield 29% more articles in public health, 26% more on gender-based violence, 25% more in gynecology, 20% more on immigrants and minorities, and 18% more on mental health. This suggests that broadening the scientific workforce may broaden the impact of science on society.

Equity for women in science cannot be achieved through a single mechanism or one actor in the complex network of science. Rather, it will take concerted effort among all stakeholders to dismantle systemic barriers to women's advancement. To that end, we present in the following tables lists of goals for several actors—individual scientists, universities and scientific organizations, science funders, professional societies and publishers, and science communicators. We also provide specific actions associated with each goal. Some of these goals focus on representation; however, taken together, they should push beyond parity toward equity.

Scientists

As noted in Chapter 7, science is a self-organizing system, in which scientists are the primary agents of change. Scientists are the producers and the consumers; the mentors, gatekeepers, and evaluators. Therefore, they have a strong obligation and opportunity to make science more equitable (Table 8.1). Collaboration is the normative mode of scientific production; thus, to achieve equity in production, scientists must restructure teams for more equitable distribution of labor (Chapter 3). As noted in Chapter 2, the percentage of early-career researchers in lead authorship positions is higher when the senior author is a woman. Teams composed by men also have lower rates of inclusion of women overall, as compared with women-led teams. Men, therefore, must take a concerted effort to engage women in research and place them in leadership positions. The professional and scientific incentives for this are clear: gender-diverse teams have higher scientific impact (Chapter 6).

Considering citations, we find that men are more likely to cite their own work and the work of other men (Chapter 6). Given that men dominate production (Chapter 1), these patterns of self-citing have significant implications for the exchange of scientific capital. Effort is necessary to ensure that literature searches are done in an inclusive manner, incorporating all relevant research. In this context, one must not minimize the effects of the ranking algorithms of retrieval systems, which often sort search results according to their numbers of citations—or similar indicators—thereby reinforcing inequalities in visibility.[5]

Scholars employed at academic institutions should also seek to incorporate a wide variety of voices into their syllabi. This simple act not only creates a more robust curriculum, it may also serve to make visible a pathway into scholarship for students traditionally underrepresented in science. Scholars should seek not to replicate themselves but to cultivate the unique talents of all their students. Furthermore, as educators, scholars also play tremendous roles as advocates and allies for their students and must provide a safe climate for students within their institution and in interactions beyond the institution.

TABLE 8.1. Summary of goals and associated actions for scientists.

Goal	Action
Acknowledge the work of women scientists.	Ensure that syllabi, seminars, and reading groups include works led by women. Cite other women in research. Following and engage with women scholars on social media. Nominate women for prestigious awards.
Provide training and mentoring for women scientists.	Include undergraduate women in research experiences. Provide women graduate students with comprehensive training experiences, including conference travel, lead authorship, team management, and grantsmanship. Create safe spaces for networking, avoiding environments that have higher chances of sexual harassment and gender discrimination for women. Engage in mentoring programs for women at all career stages: early-career researchers, midcareer researchers seeking advancement, and senior scholars working toward leadership positions. Avoid gendered language, biases, and stereotypes when writing and reading letters of recommendation, including letters for positions, promotion and tenure, and nominations for awards.
Be fair and transparent in the division of labor, authorship, and reward.	Construct authorship guidelines that are explicit in the relationship between contribution and authorship, and make these guidelines publicly available. Avoid gendered labor roles in research (such as associating certain types of labor with men or women). Conduct self-assessment of disparities and make these transparent.
Avoid gender segregation in meetings.	Listen to women in meetings and amplify their voices. Invite women to give keynotes and lectures and serve on panels. (For men, refuse to serve on "manels.") Organize workshops to expand topic areas within disciplines and create new collaborative networks that are inclusive of women.
Use research indicators responsibly.	Avoid the inappropriate use of research indicators for assessing potential team members and colleagues. Do not include research indicators without contextualization in external evaluations of scholars for promotion, tenure, or awards. Contextualize, critique, and contest indicators when they are raised in evaluation environments.

TABLE 8.1. (continued)

Goal	Action
Demonstrate zero tolerance for sexual harassment in science.	Do not discriminate by gender or harass women. Be an active bystander. Provide resources when sexual harassment and gender discrimination are reported. Do not dismiss allegations until investigated. Create and enforce codes of ethics in professional societies.
Embrace a plurality of career trajectories.	Support graduate students in multiple career trajectories; do not see replication as the goal of mentoring. Encourage students to challenge and critique established research frameworks within the discipline. Check biases while conducting evaluations of science and scientists to ensure gendered expectations are not brought into review.

As gatekeepers, researchers should ensure the representation of a diverse body of scholars at professional meetings and in the nomination to prestigious awards. One reason often given for the lack of gender representation is that "good women are hard to find." Several initiatives have focused on providing resources for conference organizers, editors, and awards committees to alleviate this concern. For example, Women in Neuroscience contains nearly 2,000 entries on women in neuroscience—about a third each of doctoral students, postdocs, and senior researchers.[6] Anne's List also focuses on the collection of names of women within subareas of neuroscience, and the Women in Cell Biology Committee of the American Society for Cell Biology maintains a list for that field.[7]

In their role as evaluators, scientists have a responsibility to use indicators responsibly and to acknowledge the inherent biases that perpetuate and reproduce inequities. They should avoid single indicators or those lacking contextualization and focus on indicators that promote social good and that lead to responsible research practices.[8] Scientists are called to remember that indicators should only be used to the extent that they reflect the values and goals of the institution.[9] An institution cannot espouse diversity as a value but continue to use indicators that systematically undervalue the labor of certain populations. Scientists are in a position to advocate for institutional change to guarantee alignment between organizational values and practices of evaluation.

Universities

Individual actors in science are largely located within the context of a scientific institution, which generally has the advantage of some autonomy in policies and practices. That is, in most countries, institutions can engage in policies—regarding hiring, promotion, tenure, and evaluation—that lead to greater equity and expand the notion of the characteristics of an ideal scientist (Chapter 7). To ensure equity in each moment of evaluation, institutions must promote organizations of equity and cultures of inclusivity (Table 8.2). They must take a firm stance on sexual harassment to ensure that the climate is hospital for all scholars to learn and work. Institutions must also provide resources and support for all scholars, contextualize performance indicators, and amplify work equitably.

As we demonstrated in Chapter 4, women are funded at lower rates than would be expected given their proportion in the workforce. They also tend to have lower rates of funding overall. University research offices should provide greater support to women scholars to encourage them to apply for funding, particularly for larger funding opportunities. Institutions can also introduce policies and practices that support women's international visibility and collaboration, acknowledging that the lower rates of international collaboration (Chapter 2) and mobility (Chapter 5) have strong implications for women's careers. Salary is another essential point of equitable reward. Several studies have noted that transparency in salary is a simple and effective way of addressing inequalities.[10] Explicit salary equity procedures are now available at most major research universities, and a growing body of literature suggests best practices for conducting salary equity reviews.[11] Institutions need to be both transparent and vigilant in addressing historical gaps and preventing the introduction of new wage inequities, within and across disciplines.

Institutions must consider not only internal practices but how they relate to the outside world. Institutional rankings are an obsession for many academic administrators. These rankings can cause distortions when they become the primary incentive for an institution and fail to align with institutional values. As a Thatcher-era adage warns, "When a measure becomes a target, it ceases to be a good measure."[12] However,

TABLE 8.2. Summary of goals and associated actions for universities.

Goal	Action
Create more inclusive promotion and tenure guidelines and processes.	Focus on content and contributions rather than indicators. Design more expansive operationalizations of excellence and provide clarity in the criteria of evaluation. Examine times to rank and promotion and provide intentional mentoring to reduce any gender disparities observed. Provide retroactive pay increases for women who extend the tenure clock for parental leave. Append implicit bias statements to invitations to external reviewers. Train promotion and tenure committees on implicit bias.
Make increases in women's authorship an institutional goal.	Monitor progress and make data public. Support the development of institutional report cards for gender equity. Participate in programs like ADVANCE, Athena SWAN, and SEA Change.
Support women scientists and provide resources for success.	Conduct salary equity reviews, implement pay transparency, and make adjustments to remove any gender disparities. Provide resources for mentoring programs for women at all career stages. Provide resources to help women obtain research grants. Provide bridge funding for early- and midcareer women and postmaternity grants for those who take parental leaves.
Take care to avoid cultural taxation when increasing representation.	Engage men as well as women in diversity initiatives. Examine and reward disproportionate teaching and service labor.
Promote women and women's work.	Increase the visibility of women in institutional communication. Provide training for women in science communication. Hire women for leadership positions. Nominate women for prestigious awards.
Reimagine the ideal worker.	Offer childcare options or stipends for graduate students and faculty with children. Provide more resources after parental leave to provide continued support for research (such as "extra hands" funding). Do not schedule meetings during times that place unnecessary burdens on parents.

(continued)

TABLE 8.2. (continued)

Goal	Action
Reimagine the ideal worker.	Provide equal opportunities for mobility and international collaborations, but do not require it for evaluation purposes.
	Do not overvalue international impact over national and local contributions.
	Allow for a mesh structure hierarchy in laboratories so that labs continue to flourish while principal investigators are on parental leave.
Support inclusive hiring practices.	Provide funding for cluster hires in fields where women are underrepresented.
	Construct equitable quotas.
	Hire women at senior levels.
	Call out gender bias in hiring practices.
	Ensure that students are taught by a diverse faculty.
Take a firm stance on sexual harassment and gender discrimination.	Implement policies that support victims of sexual harassment.
	Provide anonymized annual reports detailing statistics of recent and ongoing sexual harassment investigations, including any disciplinary actions taken.

rankings can be created so that institutions focus on topics and behavior that are targeted toward the common good, as implied in our discussion of indicators for social good (Chapter 4). Institutions are also in a position to make goals toward gender equity and provide transparency on evaluative indicators. In that context, institutions have the power to shape the behavior of individual researchers, and groups of researchers, for collective benefits.

Finally, institutions can do more to promote women in science in their communication strategies: ensuring that they are equally represented in press releases and on university promotional materials and initiating or encouraging projects such as those begun by Emily Temple-Wood (Chapter 7). They must also assess service burdens to maintain a balance between representation and cultural taxation, acknowledging that some service obligations do not have equal capital in the academic community. Women should be represented across the service-leadership spectrum, such that their labor garners appropriate reward.

Funders

Funding is both a resource and a reward. In awarding grants and prizes, councils and foundations are not only providing resources for future research; they are also rewarding scientists for past accomplishments. Funding underlies much of the reinforcement of inequities that we have described across the chapters; therefore, it must be done free from bias (Table 8.3). A key mechanism for reducing bias is the peer-review process. Funders should seek to diversify and train panels, and change evaluation criteria that implicitly favor people and past accomplishments rather than the project under evaluation. As arbiters of what and who is funded in science, funding agencies must also ensure equity at each stage of the process, in setting priorities, reviewing proposals, and creating policies for institutions that receive resources.

As described in Chapter 4, funders can exact change through mandates and requirements for funded institutions and investigators. They can also ensure that funding is provided to ensure that all can participate equally in science, through funding for childcare at conferences, extra-hands funding, and mobility programs that take contextual factors into account (Chapter 7). Finally, in monitoring, reporting, and making data available for research, they can have more empirical reflections of their own work and contribute to more global understandings of the role of funding in the scientific ecosystem.

Professional Societies and Publishers

Transparency is a critical mechanism for achieving equity in scholarly publishing. It is essential that organizations involved in such dissemination activities—from professional societies to for-profit companies—provide increased transparency around their processes and practices (Table 8.4).[13] Benchmarking exercises are useful for goal setting but also to provide comparative data for other organizations. As with funders, publishers have a responsibility to ensure that peer review is free from bias and to provide documentation of the process—either in tables of acceptance rates or by adopting practices like open peer review (see Chapter 7). Accessibility to science is mediated through ability to pay for publication:

TABLE 8.3. Summary of goals and associated actions for funders.

Goal	Action
Evaluate and fund projects, not simply people.	Avoid focusing solely or primarily on the past success of scientists when evaluating scientific proposals.
	Do not include unnecessary indicators in evaluation material (such as the Journal Impact Factor or h-indices).
	Acknowledge that evaluation of scientists over science tends to favor majority populations.
	Prevent discussions of citation indicators in panels.
Monitor and report on gender indicators.	Provide data on gender acceptance rates, funding rates, and average amounts of funding by gender.
	Practice data transparency at every level of the organization.
	Conduct regular assessments of programs to ensure gender representation in funding and topic portfolio.
Institute mandates that encourage gender equity and justice.	Require institutions to meet gender representation benchmarks, or inclusion in initiatives such as Athena SWAN, in order to receive funding.
	Mandate diversity in research teams.
	Mandate sex inclusion (or justification of exclusion) in research proposals.
	Require proposals to include authorship guidelines and mentorship plans for doctoral and postdoctoral researchers. Ensure that authorship guidelines address issues of equity in labor distribution and reward.
Reduce bias in peer review.	Invite women as reviewers, panelists, and program directors.
	Ensure diversity in the composition of peer review panels.
	Train reviewers to avoid implicit bias in evaluation.
	Fund a more diverse array of topics (prioritizing those that are disproportionately of interest to minoritized populations).
Create opportunities for greater visibility and mobility for women scientists.	Fund international collaborations and provide funding for mobility opportunities for women.
	Provide funding to allow all scholars to publish in open-access venues, following FAIR principles, to maximize the visibility of their research.
	Create funding schemes aimed at supporting early-career women.
	Provide flexible family care spending to allow women greater opportunities for mobility.

TABLE 8.4. Summary of goals and associated actions for professional societies and publishers.

Goal	Action
Increase transparency and fairness in peer review.	Provide data on the gender composition of authors, reviewers, editorial boards, and editors.
	Continually monitor and report on improvements.
	Train editorial board members and reviewers on unconscious biases.
	Assess whether acceptance and publication rates are gendered and address systematic biases where identified.
	Ensure women are represented among peer reviewers on individual articles.
	Make the content of reviews open, while maintaining the anonymity of reviewers.
Adopt inclusive publication practices.	Set article processing charges that are affordable for all authors.
	Provide early-career women with opportunities to review, without overburdening them.
	Ensure that the topical portfolio represents a range of perspectives.
	Monitor the gender of references of submitted articles and make these data public.
	Consider diversity in invited submissions as well as review papers and letters to the editor.
	Examine invitations to review for balanced gender representation.
	Allow researchers to change their names retrospectively.
Create safe spaces for women.	Develop a code of conduct for participation in conferences.
	Avoid organizing events or hosting them at venues that are unsafe for women.
	Set quotas for selection for elite speaking opportunities and awards.
Increase the visibility of women scientists.	Avoid "manels."

publishers should ensure that Article Processing Charges—which can span from several hundred to several thousand US dollars—do not present a barrier to participation in knowledge dissemination.[14] In the campaign to make science open for readers, one must be vigilant to ensure that it remains open to authors.

Professional societies—in addition to their work as publishers—also organize conferences, give awards, and sometimes act as funding agencies. In each of these roles, they must seek to increase the participation of women—for example, in invited keynotes, prestigious panels, and honorary awards. Visibility has significant effects on the selection of women for prestigious opportunities, and peripheral benefits for the representation of other women in the community.[15] For example, having at least one woman on speaker selection committees was associated with a higher proportion of women giving invited talks at the event.[16] To accomplish this, several men have made firm commitments not to serve on all-men panels (commonly named "manels"). For example, Francis Collins, the director of the National Institutes of Health in the United States, made a splash in the news when he vowed that he would no longer accept invitations to speak in forums that did not include women.[17]

Some organizations have begun transparent documentation of these inequities. BiasWatchNeuro, for example, is a website that tracks speaker composition at conferences in neuroscience, particularly with respect to gender representation.[18] It has provided a rated, interactive visualization of every conference since 2015, which shows modest improvements in gender representation. It also "watches" journals and awards. While there is no causal evidence, an analysis by *Nature* revealed that the proportion of women speakers at neuroscience conferences rose from 24% in 2011 to 42% in 2019. Other fields, however, have not observed the same effects: an examination of thirteen events in chemistry across eight years showed that the proportion of women remains constant at 24%. The article argues that "good intentions are not enough" and that "firm gender quotas or policies that compel diversity seem to reap the most success."[19]

Professional societies also have a role in creating an overall climate that is hospitable for women (as discussed in Chapter 7) and fostering the participation of researchers with a diversity of backgrounds through consideration of both the fees and location of conferences. For instance, Chapter 2 has shown that women were less likely to write single-authored pieces, especially in fields where they are uncommon. Those single-authored pieces are often published under invitation; it is important that women obtain as many opportunities to contribute as men. Inclusivity is an essential component of the move from parity to equity.

Science Communicators

Science communicators are key actors for the visibility of scientists and the dissemination of science to the public. They should seek to increase the coverage of women-led research in features, direct quotes, and the science they report; and they should monitor such coverage (Table 8.5). Science communicators must consider the way in which they present women, from language to visual representation, to ensure that they are not reinforcing gendered roles in science (see Chapter 7). They may also consider the practices of interviewing and whether the times and modalities create barriers to participation. Science communication should also engage in reflexivity about its own history of gender bias, including harassment, salary inequities, and gender disparities in assignments (for example, men are much more likely to report on technology and general science than women).[20] Equity in science communication requires the field to consider its own labor practices and the gendered way in which it reflects scientific labor.

TABLE 8.5. Summary of goals and associated actions for science communicators.

Goal	Action
Amplify women scientists and their work.	Increase coverage of women in direct quotes, features, and representation in advertisements.
	Ensure that visual representations of women highlight a diversity of career trajectories.
	Avoid gendered language in describing women scientists.
	Be critical of article titles or quotes that inflate accomplishments of scientists.
	In the case of papers with multiple authors, try to target the women of the team.
	Monitor and report on gendered coverage.
	Cover a diverse range of topics and disciplines.
	Use asynchronous means of interviewing, acknowledging that women may have domestic duties that constrain their time and ability to do a synchronous interview.
Provide opportunities and training for women in science communication.	Seek gender representation in professional opportunities.
	Provide training programs for women scientists in science communication.

Conclusion

Transformations of the scientific system—toward one that is open to and benefits all members of society—requires informed and collective action. It will take concerted effort to share data and best practices and work intentionally toward meaningful structural change. We call on the scientific community to engage with these actions *as scientists:* to design and assess interventions, document success and failure, and communicate the results of these actions openly and widely. The actions toward data collection and transparency are critical to move the community toward an evidenced-based scientific system. It is not, however, mere data collection but a change in the labor practices and cultural and institutional systems of science that is required.

Change also requires humility and reflexivity, to acknowledge failure and to experiment with new systems that disrupt the status quo. The vignettes presented in this book demonstrate the deeply rooted and systemic nature of disparities; the empirical results provide evidence that strong differences in scientific labor and reward remain. The notable disparities across all areas of scientific labor—from production to mobility—demonstrate the cycles of reproduction that perpetuate barriers. Disruption of these cycles calls for collective action from all those who labor in and for science.

APPENDIX: MATERIALS AND METHODS

NOTES

ACKNOWLEDGMENTS

INDEX

Appendix

MATERIALS AND METHODS

Bibliometric Data

Bibliometric methods were developed in the early twentieth century to help librarians manage their scientific collections. The creation of the Science Citation Index in the early 1960s by chemist–turned–information scientist Eugene Garfield led to a diversity of uses of those methods, from research evaluation to sociological and historical studies of science. Bibliometric data for this book are drawn from Clarivate Analytics' Web of Science (WoS)—the successor of the Science Citation Index—for the 2008–2020 period. WoS and Scopus (produced by Elsevier) are considered the standard databases for bibliometric analyses, due to their highly curated data and wide coverage.[1] However, there are several caveats: WoS has low coverage of literature in languages other than English, as well as a lack of coverage of nonjournal literature, which leads to lower representation of publication data in the social sciences and humanities and in disciplines such as computer science and engineering.[2] Therefore, through our analysis of the WoS, the bibliometric data presented in this book are a stronger reflection of research outputs—and of the various forms the gender gap takes in those publications—for disciplines of natural sciences and medical sciences, as well as for research from English-speaking countries. However, it reasonable to believe that the disparities observed in journal articles are also observed in books and conference proceedings. In terms of publication language, evidence from Québec researchers has shown that the gaps observed in English-language publications were also observed for French-language publications.[3]

The first year covered in the data (2008) was selected because it was the year in which Clarivate Analytics started to index links between authors and their institutions, which allows for the creation of a country-level gender assignment algorithm, based on given names that were indexed from 2006 onward. For the 2008–2020 period, Clarivate's three main citation indices—Science Citation Index Expanded, Social Science Citation Index, and Arts & Humanities Citation Index—indexed 27,462,105 scholarly documents, of which 20,085,758 (73.1%) are articles and reviews, the two document types covered in our analysis. Unlike other types of documents published by journals—such as letters to the editor, editorials, and news items—articles and reviews are generally considered original contributions to knowledge and have been evaluated through peer review.

The discipline and specialty classification used for the bibliometric analyses is that developed by the firm CHI Research (now the Patent Board) for the US National Science Foundation (NSF).[4] The main advantages of this classification scheme over that provided by the WoS are that it has a two-level classification (discipline and specialty)—which allows the use of two different levels of aggregation—and that it categorizes each journal into only one discipline and specialty, which prevents double counts of papers when they are assigned to more than one discipline. The NSF classification scheme, however, does not provide any disciplinary classification for journals in the arts and humanities—mainly journals indexed by the Arts and Humanities Citation Index. Hence, these journals were manually categorized by the team of the Observatoire des sciences et des technologies—the licensee of the WoS database used in this book—in specialties of the arts and the humanities. The list of countries and regions used is that of the United Nations under the standard M49.[5] As the WoS uses specific country classifications—especially for the United Kingdom, which is divided into England, North Ireland, Scotland, and Wales—we grouped those countries into the United Nations categorization. Similarly, the WoS categorizes Taiwan as a specific entity; we kept this categorization in our analysis.

Gender Assignment

The gender assignment algorithm used throughout this book has been previously described by the authors and colleagues.[6] It was developed using the US Census lists of names and gender, along with various general lists from Wikipedia and country-specific lists.[7] We acknowledge the limitation of algorithmic gender assignments, which only consider gender in a binary manner. Other genders, unfortunately, can only be obtained from self-identification and cannot be assigned algorithmically. The US Census list was applied to all countries, except those for which country- or region-level lists were obtained. Given that several names in the US Census—such as Val, Britt, or Leslie—could be assigned to either a man or a woman, a name had to be ten times more commonly assigned to a given gender to be assigned to that gender. For example, the name Ariel, which is as common for men and for women, was categorized as unknown. Additional gender assignment using family names was also performed for Slavic countries. Men's family names typically end in *-ov, -ev,* or *-in,* while women's end in *-ova, -eva,* or *-ina.* These suffixes were thus applied to Russian authors, as well as to authors from other countries where 95% or more of the women's or men's names already assigned ended in one of the suffixes (Czechia, Bulgaria, Latvia, Kazakhstan, Uzbekistan, Lithuania, and Luxembourg).

The accuracy of the algorithm was originally assessed using a sample of 5,000 records of individual authors: 1,000 categorized as men, 1,000 categorized as women, 1,000 with initials, and 2,000 unknowns. Using their full name, country, institution, and email address, we traced these individuals online and verified their declared or perceived gender. Globally, 98.3% of those assigned as men in the algorithm presented as men, while 87% of those assigned as women presented as women. We also observed a distribution of 24% women and 76% men for initials, as well as a distribution of 34% women and 66% men for the unknown names.[8] The algorithm has been updated since this time; however, the initial validation results suggest that the algorithm may slightly overestimate the percentage of women in authorship lists.

We also performed another round of validation using the genders declared in the contributorship survey presented in Chapter 3 (Table A.1). Quite strikingly, the same percentage (77.0%) of both men and women—as declared in the survey—were correctly categorized according to our gender assignment algorithm. The percentage of women categorized as unknown (17.1%) was slightly lower than that of men (19.0%). Women were also slightly more likely to be assigned as men (5.9%) than the opposite (men assigned as women—4.0%). Both transgender individuals and those of other genders were, roughly, as likely to be categorized as women or men, while the vast majority of those who have preferred not to answer that question were categorized as men.

TABLE A.1.   Declared gender of survey respondents, as a function of their algorithmically assigned gender.

Declared gender	Algorithmic gender							
	Women		*Men*		*Unknown*		*All respondents*	
	N	%	N	%	N	%	N	%
Women	2,713	77.0	207	5.9	604	17.1	3,524	100.0
Men	339	4.0	6,566	77.0	1,622	19.0	8,527	100.0
Transgender	3	42.9	4	57.1	–	0.0	7	100.0
Other	3	60.0	1	20.0	1	20.0	5	100.0
Prefer not to answer	14	14.0	64	64.0	22	22.0	100	100.0
Total	3,072	25.3	6,842	56.3	2,249	18.5	12,163	100.0

Note: The numbers are greater than those analyzed in Chapter 3, as we are including here respondents who did not identify as men or women.

Figure A.1 presents the average of fractionalized authorship—that is, the average percentage of authors within papers—to which a gender could be assigned, by country. At the world level, our algorithm managed to assign a gender to 74.4% of fractionalized authorships. This percentage, however, varies by country: given the use of family names for Russia and other countries with Slavic names, the coverage is much higher, with the quasi-totality of authorships being assigned a gender. Australia, Canada, New Zealand, and the United States exhibit a higher-than-average percentage of authorships assigned to a gender, which can partly be attributed to the use of sources such as the US Census and Wikipedia. China and Japan, as well as Scandinavian countries and most countries in the western part of South America, are also better covered. On the other hand, assignment of authorships from most western European countries is slightly below average. It is also below average for African countries, Middle Eastern countries, Southeast Asian countries, and India.

Figure A.2 provides, by discipline, the percentage of fractionalized authorships, by gender as well as for initials and unknowns. Disciplines vary in their proportion of authorships to which a gender could be assigned, and this difference is mostly due to the use of initials. More specifically, disciplines have a very similar percentage of unknowns, which range from a minimum of 9% in psychology to 17% in arts. However, the percentage of authorships with initials varies sizably, from 0%–1% in arts and humanities to 20% in clinical medicine and 25% in physics.

FIGURE A.1. Percentage of fractionalized authorships to which a gender could be assigned, by country. Parts of the world with no data (Antarctica) are not shown, 2008–2020.

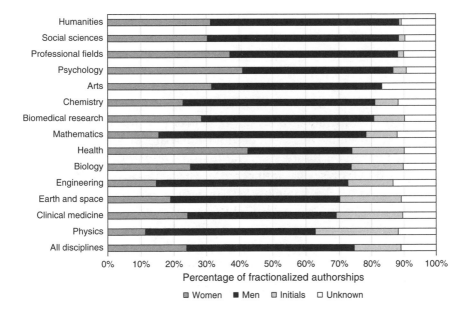

FIGURE A.2. Percentage of fractionalized authorships to which a gender could be assigned, by discipline, 2008–2020.

As a result, the percentage of fractionalized authorships to which a gender could be assigned is above 88% in humanities and social sciences, but 63% in physics.

World Bank Data

Country-level data from the World Bank were downloaded in Excel format in May 2021.[9] The dataset contained 922 indicators related to gender and economic power, education, health, public life, and decision making, agency, and inequalities.[10] While the data cover 1960–2019—with a large share of missing data for some countries—we focused our analysis on the 2008–2017 period to cover for publication lags. The data for each country and indicator were averaged for the entire period to be comparable with the bibliometric data that covers the 2008–2020 period. These averages excluded the years where some data were missing. Data were matched with countries' names as they appeared in the United Nations' list of countries.

Data on Contributorship from the Public Library of Science

Launched in 2014, the CRediT (Contributor Roles Taxonomy) categorizes contributions made to scholarly papers into fourteen categories (Table A.2). Given the increased need for transparency in the attribution of credit and responsibility,[11] several journals *(eLife, Cell, F1000)* and publishers (PLOS, Elsevier, Springer, BMJ) have adopted the taxonomy. Our analysis is based on one of these publishers—the Public Library of Science (PLOS)—which provided all its contributorship information for papers published between June 15th 2017 and 31st of December 2018. The data included metadata for all papers published in PLOS journals, including their publication date, Digital Object Identifier (DOI), journal name, author name as it appears on the paper, and each of the associated CRediT contributions.

Table A.3 presents the characteristics of the dataset. It shows, unsurprisingly, that the bulk of contributions are in papers published in the megajournal *PLOS ONE* (87.9%), and only a few others in *PLOS Biology*.[12] Important differences are observed in terms of mean number of

TABLE A.2. CRediT contributions and their associated definitions.

Contribution	Definition
Conceptualization	Ideas; formulation or evolution of overarching research goals and aims.
Data curation	Management activities to annotate (produce metadata), scrub data, and maintain research data (including software code, where it is necessary for interpreting the data itself) for initial use and later reuse.
Formal analysis	Application of statistical, mathematical, computational, or other formal techniques to analyse or synthesize study data.
Funding acquisition	Acquisition of the financial support for the project leading to this publication.
Investigation	Conducting a research and investigation process, specifically performing the experiments, or data/evidence collection.
Methodology	Development or design of methodology; creation of models.
Project administration	Management and coordination responsibility for the research activity planning and execution.
Resources	Provision of study materials, reagents, materials, patients, laboratory samples, animals, instrumentation, computing resources, or other analysis tools.
Software	Programming, software development designing computer programs; implementation of the computer code and supporting algorithms; testing of existing code components.
Supervision	Oversight and leadership responsibility for the research activity planning and execution, including mentorship external to the core team.
Validation	Verification, whether as a part of the activity or separate, of the overall replication/reproducibility of results/experiments and other research outputs.
Visualization	Preparation, creation and/or presentation of the published work, specifically visualization/data presentation.
Writing—original draft	Preparation, creation and/or presentation of the published work, specifically writing the initial draft (including substantive translation).
Writing—review and editing	Preparation, creation and/or presentation of the published work by those from the original research group, specifically critical review, commentary or revision—including pre- or post-publication stages.

TABLE A.3. Number of papers published with CRediT contributions, mean number of authors, and mean number of CRediT contributions per paper, by PLOS journal.

Journal	Number of papers	Mean number of authors	Mean number of contributions
PLOS Biology	13	7.2	11.85
PLOS Computational Biology	763	4.9	11.07
PLOS Genetics	786	8.5	11.09
PLOS Medicine	250	14.2	10.84
PLOS Neglected Tropical Diseases	1,144	9.1	11.15
PLOS ONE	27,057	6.8	10.57
PLOS Pathogens	757	9.4	10.97
All journals	30,770	7.0	10.63

authors per paper, with *PLOS Computational Biology* seeing on average slightly less than five authors per paper, while *PLOS Medicine* has almost three times that. Mean number of contributions is, however, quite constant across journals, with a maximum of 11.8 in *PLOS Biology* and a minimum of 10.6 in *PLOS ONE*.

Contribution information provided by PLOS did not, however, contain author order. To obtain this information, we had to match each PLOS paper with its record in our in-house version of Clarivate Analytics' WoS based on the DOI; this was feasible for 30,054 papers (97.7% of the PLOS dataset), which totaled 222,938 authorships. Once the papers were matched with the WoS, we matched each author in both data sources to obtain their individual order in the authors' list. This was first based on the full name string—for example, Cassidy Rose Sugimoto = Cassidy Rose Sugimoto. However, several names could not be matched because they were written in different manners in both databases (Cassidy R. Sugimoto, Cassidy Sugimoto); therefore, we performed additional matching focusing on specific parts of the name string. More specifically, we iteratively focused on the first and last two to five characters of the names. This allowed us to match 221,637 authorships (99.4% of the sample).

For authors who could be assigned an author order, we also assigned a gender based on their given names using the algorithm developed by Larivière and colleagues.[13] The algorithm assigned a gender to 82.2% of the sample (Table A.4)—a percentage that is above the assignation performed

TABLE A.4. Percentage of authorships assigned to a gender, by author order.

Gender	First N	First %	Middle N	Middle %	Last N	Last %	Any order N	Any order %
Gender assigned	26,005	79.9	129,198	82.3	27,064	84.4	182,267	82.2
Women	12,094	37.2	52,106	33.2	8,600	26.8	72,800	32.8
Men	13,911	42.7	77,092	49.1	18,464	57.6	109,467	49.4
Initials	382	1.2	2,085	1.3	447	1.4	2,914	1.3
Unisex	704	2.2	3,899	2.5	911	2.8	5,514	2.5
Unknown	5,462	16.8	21,847	13.9	3,633	11.3	30,942	14.0
Total	32,553	100.0	157,029	100.0	32,055	100.0	221,637	100.0

on the WoS. This percentage varies by author order, however, with a higher proportion of last authors assigned a gender, and a lower proportion of first authors. The percentage of women authorships in the PLOS dataset represents 39.9% of authorships to which a gender could be assigned, which is slightly greater than the percentage of women authorships found in the WoS for disciplines of the medical sciences (about 35%).

Survey on Authors' Contributions

As the PLOS dataset previously used in contributorship studies is heavily biased toward medical sciences, we conducted a survey of corresponding authors across all fields to better understand the tasks performed by men and women authors.[14] To this end, we created a sample of corresponding authors from articles published in 2016 and indexed in the WoS. For each corresponding author surveyed, we asked them to list the contributions made by each of the coauthors of the paper (including themselves). These contributions were the following:

- Design
- Data collection (including historical and survey research)
- Performing experiments (including survey-based experimentation)
- Analysis/interpretation
- Writing the manuscript
- Contribution of resources (financial, equipment, reagents)
- Other, please specify: _____
- No contribution
- Unsure

To obtain results for disciplines outside medical sciences, we created a stratified random sample of 132,917 corresponding authors, which represented about 11.5% of all papers from each discipline—ranging from 9.8% of papers in chemistry to 19.9% in mathematics. We restricted the analysis to papers between three and ten authors to focus on papers that contain collaboration—and therefore division of labor—but also that did not include hyperauthorship, to make it feasible for corresponding authors to remember the contribution each author performed. Restricting to these collaborative papers reduced the set of analyzed papers from 2,002,939 to 1,158,590 (57.8% of the dataset). This percentage varies greatly by discipline, however: while 80.3% of papers in chemistry fall in this category, this percentage is 22.3% in the social sciences (Table A.5).

We obtained 12,001 completed surveys, for a global response rate of 9.0%. The response rate was higher in some disciplines than others: while only 6.7% and 7.1% of corresponding authors from, respectively, clini-

TABLE A.5. Number of respondents, surveyed corresponding authors, number of papers with three to ten authors and all papers, 2016.

Discipline	Responded		Surveyed		2016 WoS papers with three to ten authors		All 2016 WoS
	N	%	N	%	N	%	
Biology	1,223	11.8	10,395	12.0	86,925	71.4	121,690
Biomedical research	1,336	8.4	15,871	10.3	154,124	65.4	235,802
Chemistry	884	7.1	12,444	9.8	127,307	80.3	158,567
Clinical medicine	2,299	6.7	34,280	10.2	335,252	48.0	698,007
Earth and space	1,097	12.3	8,922	12.1	73,940	70.8	104,483
Engineering and technology	2,093	8.1	25,823	13.2	195,813	71.6	273,565
Health	380	12.0	3,158	11.2	28,268	45.5	62,087
Mathematics	388	10.1	3,834	19.9	19,228	37.9	50,765
Physics	865	9.1	9,540	11.8	80,892	62.4	129,665
Professional fields	538	15.8	3,404	17.7	19,219	35.0	54,903
Psychology	387	15.5	2,491	11.6	21,545	52.1	41,340
Social sciences	511	18.5	2,755	17.1	16,077	22.3	72,065
All disciplines	12,001	9.0	132,917	11.5	1,158,590	57.8	2,002,939

TABLE A.6. Number and percentage of respondents by discipline and gender.

Discipline	Women		Men		Other		Prefer not to answer	
	N	%	N	%	N	%	N	%
Biology	427	34.9	779	63.7	1	0.1	16	1.3
Biomedical research	420	31.4	892	66.8			24	1.8
Chemistry	183	20.7	682	77.1	2	0.2	17	1.9
Clinical medicine	830	36.1	1,431	62.2	2	0.1	36	1.6
Earth and space	301	27.4	777	70.8	1	0.1	18	1.6
Engineering and technology	314	15.0	1,735	82.9	2	0.1	42	2.0
Health	223	58.7	152	40.0			5	1.3
Mathematics	78	20.1	299	77.1	1	0.3	10	2.6
Physics	119	13.8	729	84.3	2	0.2	15	1.7
Professional fields	195	36.2	336	62.5			7	1.3
Psychology	192	49.6	192	49.6			3	0.8
Social sciences	188	36.8	316	61.8	1	0.2	6	1.2
All disciplines	3,470	28.9	8,320	69.3	12	0.1	199	1.7

cal medicine and chemistry participated in the survey, this percentage was more than twice as high for professional fields, psychology, and social sciences. While this imbalance may seem problematic, it leads to better knowledge of how labor is divided across gender in domains where contributions have been invisible and, therefore, understudied. Table A.6 presents the gender distribution of survey respondents, by discipline.

Academic Analytics

The data from Academic Analytics are used to assess gender differences at the level of individual researchers' funding portfolios. Based in the United States, Academic Analytics provides scientometric indicators to university administrators on funding, awards, and publications of individual faculty members and departments. Although the details of the data collection process remain proprietary, the data are obtained in collaboration with research institutions as well as from publicly available sources such as Crossref, federal funding agencies, and the institutional websites of universities.

Our analysis is based on the 2017 data release, which contains scientometric indicators for 165,666 tenured and tenure-track faculty located in

399 universities in the United States. In addition to sociodemographic data on faculty (such as year of PhD), the dataset provides bibliometric indicators on numbers of papers over the last five years and their citations, as well as information on grants received over the same period. It is worth noting that the database does not provide records on the specific awards or funding received, or on the publications. Each row is a faculty member, for which various counts of awards, funding, or publications are provided.

The dataset uses a hierarchical three-tiered disciplinary taxonomy based on departmental affiliations. At its lowest level of aggregation, the classification contains 171 distinct specialties. Those specialties were mapped to the fourteen broad categories from the NSF used in the rest of this book. The dataset contains a fair number of individual researchers ($N = 42,500$) who have joint appointments or are affiliated with several disciplinary domains. Those cases were duplicated—that is, counted more than once—for each of the disciplines to which they are assigned. Gender was assigned using the same methodology presented earlier. Of all the faculty indexed, we managed to assign a gender to 95.0% (32.9% women and 62.1% men), and 5% were unknown.

National Funders

Data for the NSF were downloaded in November 2021.[15] Each XML file was converted and merged into one Excel file for data compilation. This file contains 442,745 awards between 1980 and 2022; we only kept data between 1980 and 2020 for completeness ($N = 430,204$), which totals US\$177,317,765,899 in funding. The database included fifteen different types of awards; only continuing grants and standard grants were analyzed in the grant category (they account for 68.6% of all the funding), and fellowship awards and fellowships were considered in the fellowship category (2.9% of all funding). The rest of the funding falls into cooperative agreements (22.7%) as well as various forms of contracts (5.8%). Information on directorates was directly provided in the database. Gender was assigned using the same method as presented earlier; the principal investigators responsible for 89.4% of the total funding awarded were assigned a gender (18.2% women and 71.4% men), while 89.3% of projects' principal investigators were assigned a gender.

Data for Canadian funders was obtained through the tri-council data cube created by the Observatoire des sciences et des technologies, which merges federal funding sources—Canadian Institutes of Health Research, the Natural Sciences and Engineering Research Council, the Social Sciences and Humanities Research Council, and the Canada Research Chairs program—into one combined file in order to perform global analyses of funding trends. More specifically, the combined dataset aligns the various program categories and makes sure the data from the three councils are homogeneous. The combined dataset contains 359,618 grants and scholarships, for a total of Can$39,034,321,118 (41.8% from the Canadian Institutes of Health Research, 44.7% from the Natural Sciences and Engineering Research Council, and 13.5% from the Social Sciences and Humanities Research Council). Gender was assigned using the same method as presented earlier; 92.7% of funds were assigned a gender (25.9% women and 66.8% men), while 91.3% of projects were assigned a gender.

Citation Diversity Statement

Citation Diversity Statements have been advanced in recent years as a practice to provide transparency and mitigate citation biases.[16] We engaged in this practice at the final stages of proofreading the manuscript. Our method involved manual coding of the first authors of the 442 unique references cited in the book. Fifty-five of the references were anonymous (e.g., organizational entities or unnamed editors). Of the remaining 387 references (representing all document types), 52 percent were first-authored by a scholar identifying as a woman and 48 percent by a man. Our method is limited by our personal knowledge of scholars' identities and the information conveyed in public websites (e.g., pronouns). We recognize that this is likely to under-represent or misclassify nonbinary, transgender, intersex, and other gender identities beyond the binary. We note, however, that our representation of women authors exceeds what would be expected in the general pool of authors (Chapter 1).

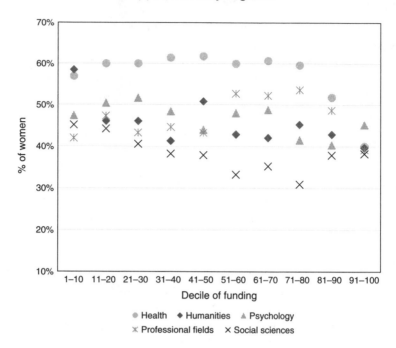

FIGURE A.3. Percentage of women researchers, by decile of funding and NSF discipline of the social sciences and humanities, 2012–2016. Academic Analytics Database.

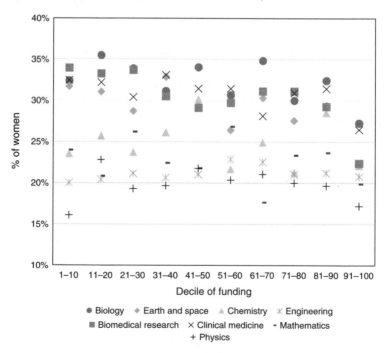

FIGURE A.4. Percentage of women researchers, by decile of funding and NSF discipline of the natural and medical sciences, 2012–2016. Academic Analytics Database.

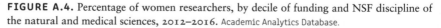

NOTES

Introduction

1. Londa Schiebinger, *The Mind Has No Sex? Women in the Origins of Modern Science* (Cambridge, MA: Harvard University Press, 1991).

2. Angela Saini, *Inferior: How Science Got Women Wrong—and the New Research That's Rewriting the Story* (Boston: Beacon, 2017); Vesna Crnjanski Petrovich, "Women and the Paris Academy of Sciences," *Eighteenth-Century Studies* 32, no. 3 (1999): 383–390.

3. Increases in participation have not been linear. For example, in the United States, the proportion of women receiving doctoral degrees was higher in the 1920s and 1930s than it was in the 1940s through the 1960s. It only recovered to the rates of the 1930s in the 1970s.

4. World Economic Forum, *Global Gender Gap Report 2021* (Geneva: World Economic Forum, 2021), https://www3.weforum.org/docs/WEF_GGGR_2021.pdf.

5. "Gender, Institutions and Development Database (GID-DB) 2019," OECD, accessed February 13, 2022, https://stats.oecd.org/Index.aspx?DataSetCode=GIDDB2019.

6. M. F. Fox, "Gender, Hierarchy, and Science," in *Handbook of the Sociology of Gender,* ed. Janet Saltzman Chafetz (Boston: Springer, 2006), 453.

7. M. Beard, *Women and Power: A Manifesto* (New York: Liveright, 2017), 84–85.

8. Asaf Levanon, Paula England, and Paul Allison, "Occupational Feminization and Pay: Assessing Causal Dynamics Using 1950–2000 U.S. Census Data," *Social Forces* 88, no. 2 (2009): 865–891.

9. Claire Cain Miller, "As Women Take Over a Male-Dominated Field, the Pay Drops," *New York Times,* March 18, 2016.

10. B. F. Reskin and P. A. Roos, *Job Queues, Gender Queues: Explaining Women's Inroads into Male Occupations* (Philadelphia: Temple University Press, 2009).

11. Robert K. Merton, "The Normative Structure of Science," in *The Sociology of Science: Theoretical and Empirical Investigations* (Chicago: University of Chicago Press, 1973), 267–278.

12. Fox, "Gender, Hierarchy, and Science," 441, 444.

13. S. Spencer, *French Women and the Age of Enlightenment* (Bloomington: Indiana University Press, 1984).

14. Margaret W. Rossiter, *Women Scientists in America,* vol. 1, *Struggles and Strategies to 1940* (Baltimore: Johns Hopkins University Press, 1982).

15. Sydney Ross, "*Scientist:* The Story of a Word," *Annals of Science* 18, no. 2 (1962): 65–85.

16. Melinda Baldwin, *Making "Nature": The History of a Scientific Journal* (Chicago: University of Chicago Press, 2015).

17. M. Holloway, "Profile: Ruth Hubbard—Turning the Inside Out," *Scientific American* 272, no. 6 (1995): 49–50.

18. Ruth Hubbard, "Reflections on the Story of the Double Helix," *Women's Studies International Quarterly* 2, no. 3 (1979): 261–273.

19. Throughout this book, we use *gender* to refer to the inferred or declared social identities of scientists, while we use *sex* only in the few instances where we are discussing biological differences. We extend this practice in our utilization of *men* and *women* throughout the book to refer to gender and *male* or *female* for instances where we are referring to sex.

20. V. Regitz-Zagrosek, "Sex and Gender Differences in Health," *EMBO Reports* 13, no. 7 (2012): 596–603.

21. Cassidy R. Sugimoto et al., "Factors Affecting Sex-Related Reporting in Medical Research: A Cross-Disciplinary Bibliometric Analysis," *Lancet* 393, no. 10171 (February 2019): 550–559.

22. Angela Saini, *Inferior: How Science Got Women Wrong-and the New Research That's Rewriting the Story* (Boston: Beacon, 2017), 10.

23. Similar predictions have been made by others, using different datasets. For example, using the preprint server arXiv, it was estimated that the gender ratio of senior physicists will come within 5% of parity by the year 2276. Luke Holman, Devi Stuart-Fox, and Cindy E. Hauser, "The Gender Gap in Science: How Long Until Women Are Equally Represented?," *PLOS Biology* 16, no. 4 (2018): e2004956.

24. M. F. Fox, K. Whittington, and M. Linkova, "Gender, (In)equity, and the Scientific Workforce," in *The Handbook of Science and Technology Studies,* 4th ed., ed. Ulrike Felt et al. (Cambridge, MA: MIT Press, 2017), 701–731.

25. Caroline Criado-Perez, "The Deadly Truth about a World Built for Men—from Stab Vests to Car Crashes," *Guardian,* February 23, 2019.

26. Eva Botkin-Kowacki, "NASA's Spacesuits Have a Gender Problem. These Women Are Fixing It," *Christian Science Monitor,* July 20, 2020.

27. In fact, the ambiguity of scientific labor within academe has been a point of contention in some countries. For example, in the United States, data released from a public institution on salaries demonstrated that there was an inverse relationship between teaching loads and salaries. This led to outcry from some, suggesting that labor was not being adequately compensated at these institutions. This public debate demonstrated the strong public opinion that associated academic labor with teaching obligations rather than research activity. Richard Vedder, "Teaching Loads and Affordability: The University of Texas Data," *Chronicle of Higher Education,* May 23, 2011, https://www.chronicle.com/blogs/innovations/teaching-loads-and-affordability-the-university-of-texas-data-2/29469.

28. Cassidy R. Sugimoto and Vincent Larivière, *Measuring Research: What Everyone Needs to Know* (New York: Oxford University Press, 2018).

29. Andrew Tsou, Jutta Schickore, and Cassidy R. Sugimoto, "Unpublishable Research: Examining and Organizing the 'File Drawer,'" *Learned Publishing* 27, no. 4 (2014): 253–267.

30. Vincent Larivière et al., "Vanishing Industries and the Rising Monopoly of Universities in Published Research," *PLOS ONE* 13, no. 8 (2018): e0202120.

31. Vincent Larivière et al., "The Place of Serials in Referencing Practices: Comparing Natural Sciences and Engineering with Social Sciences and Humanities," *Journal of the American Society for Information Science and Technology* 57, no. 8 (2006): 997–1004.

32. C. R. Sugimoto et al., "The Academic Advantage: Gender Disparities in Patenting," *PLOS ONE* 10, no. 5 (2015): e0128000; L. L. Wang et al., "Gender Trends in Computer Science Authorship," *Communications of the ACM* 64, no. 3 (2021): 78–84.

33. D. J. de Solla Price, "The Science of Science," in *The Science of Science: Society in a Technological Age,* ed. M. Goldsmith and A. MacKay (London: Scientific Book Club, 1964), 202.

34. R. Rousseau, "Naukometriya, Nalimov and Mul'chenko," *COLLNET Journal of Scientometrics and Information Management* 15, no. 1 (2021): 213–224.

35. De Solla Price, "Science of Science," 200–201.

36. Blaise Cronin, "Hyperauthorship: A Postmodern Perversion or Evidence of a Structural Shift in Scholarly Communication Practices?," *Journal of the American Society for Information Science and Technology* 52, no. 7 (2001): 558–569.

37. F. Karimi et al., "Inferring Gender from Names on the Web: A Comparative Evaluation of Gender Detection Methods," in *Proceedings of the 25th International Conference Companion on World Wide Web* (New York: Association for Computing Machinery, April 2016), 53–54.

38. E. Caron and N. J. van Eck, "Large Scale Author Name Disambiguation Using Rule-Based Scoring and Clustering," in *Proceedings of the 19th International Conference on Science and Technology Indicators* (Leiden: Center for Science and Technology Studies, Leiden University, 2014), 79–86.

39. E. Smith et al., "Misconduct and Misbehavior Related to Authorship Disagreements in Collaborative Science," *Science and Engineering Ethics* 26, no. 4 (2020): 1967–1993.

40. P. Mongeon and A. Paul-Hus, "The Journal Coverage of Web of Science and Scopus: A Comparative Analysis," *Scientometrics* 106, no. 1 (2016): 213–228; V. Larivière et al., "The Place of Serials in Referencing Practices: Comparing Natural Sciences and Engineering with Social Sciences and Humanities," *Journal of the American Society for Information Science and Technology* 57, no. 8 (2006): 997–1004; C. Lisée, V. Larivière, and É. Archambault, "Conference Proceedings as a Source of Scientific Information: A Bibliometric Analysis," *Journal of the American Society for Information Science and Technology* 59, no. 11 (2008): 1776–1784.

41. Margaret W. Rossiter, *Women Scientists in America,* 3 vols. (Baltimore: Johns Hopkins University Press, 1982–2012).

42. We refer the reader to several comprehensive biographies of these women in the relevant chapters.

1. Production

1. "Most Influential Women in British Science History," Royal Society, accessed May 24, 2021, https://royalsociety.org/topics-policy/diversity-in-science/influential-british-women -science/.

2. Melinda Baldwin, *Making "Nature": The History of a Scientific Journal* (Chicago: University of Chicago Press, 2015).

3. Marlene Rayner-Canham and Geoff Rayner-Canham, "Pounding on the Doors: The Fight for Acceptance of British Women Chemists," *Bulletin for the History of Chemistry* 28, no. 2 (2003): 110–119.

4. "Women and the Fellowship of the Chemical Society," *Nature* 78 (July 1908): 226.

5. Henry E. Armstrong, "Mrs. Hertha Ayrton," *Nature* 112 (December 1923): 801.

6. Baldwin, *Making "Nature,"* 4.

7. Margaret W. Rossiter, *Women Scientists in America,* vol. 1, *Struggles and Strategies to 1940* (Baltimore: Johns Hopkins University Press, 1982), 17.

8. Angela Saini, *Inferior: How Science Got Women Wrong—and the New Research That's Rewriting the Story* (Boston: Beacon, 2017).

9. Rossiter, *Women Scientists in America,* 15, 51, 16, 60, 53.

10. Cynthia Cockburn, *Machinery of Dominance: Women, Men, and Technical Know-How* (Boston: Northeastern University Press, 1988); Anne Witz, *Professions and Patriarchy* (New York: Routledge, 1992).

11. Julie Des Jardins, *The Madame Curie Complex: The Hidden History of Women in Science* (New York: Feminist Press at the City University of New York, 2010).

12. That is, names on the bylines of academic articles. Fractionalized authorships are defined as the average percentage of authors on a given paper who are of a given gender. For instance, a set of two papers, one with ten authors (five female) and one with four authors (one female), would have a female fractionalized authorship of 38.5%.

13. This classification categorizes each journal indexed in the Web of Science into one of fourteen disciplines and 143 specialties. See K. Hamilton, *Subfield and Level Classification of Journals,* no. 2012-R (Cherry Hill: CHI Research, 2003).

14. These ten specialties account for a mere 4% of all papers published from 2008 to 2018.

15. B. F. Reskin and P. A. Roos, *Job Queues, Gender Queues: Explaining Women's Inroads into Male Occupations* (Philadelphia: Temple University Press, 2009).

16. Asaf Levanon, Paula England, and Paul Allison, "Occupational Feminization and Pay: Assessing Causal Dynamics Using 1950–2000 U.S. Census Data," *Social Forces* 88, no. 2 (2009): 865–891.

17. Ruth Oldenziel, *Making Technology Masculine: Men, Women, and Modern Machines in America, 1870–1945* (Amsterdam: Amsterdam University Press, 1999).

18. It is worth noting that, among the disciplines with the lowest proportion of women (such as physics and engineering), there is a higher share of women in applied specialties (such as chemical physics and chemical engineering). Although chemistry has a slightly higher percentage of women authorships overall, we also observe a higher percentage of women in applied chemistry.

19. Sarah-Jane Leslie et al., "Expectations of Brilliance Underlie Gender Distributions across Academic Disciplines," *Science* 347, no. 6219 (January 2015): 262–265.

20. Debbie Ma et al., "21% versus 79%: Explaining Philosophy's Gender Disparities with Stereotyping and Identification," *Philosophical Psychology* 31, no. 1 (2018): 68–88.

21. "Economics Is Uncovering Its Gender Problem," *Economist,* March 21, 2019.

22. Catherine Riegle-Crumb and Chelsea Moore, "The Gender Gap in High School Physics: Considering the Context of Local Communities," *Social Science Quarterly* 95, no. 1 (April 2013): 253–268.

23. G. Sonnert, M. F. Fox, and K. Adkins, "Undergraduate Women in Science and Engineering: Effects of Faculty, Fields, and Institutions over Time," *Social Science Quarterly* 88, no. 5 (2007): 1333–1356.

24. Mary Frank Fox, "Gender, Family Characteristics, and Publication Productivity among Scientists," *Social Studies of Science* 35, no. 1 (February 2005): 131.

25. The year 2008 was the first in which the Web of Science made a link between authors and addresses. Combined with the indexing of given names of authors in 2006, this made it possible to perform country-level assignments of genders to authors from different countries.

26. This is exemplified by the fact that, when fractional counting is used—by allocating to each author a fraction of the number of authors on the paper—the gender gap in research production is smaller (16.1% vs. 26.8%) than with full counting.

27. In total, we identified 1,393,635 distinct researchers; we assigned a gender to 62.4% of these researchers. However, a large proportion of these authors wrote only a single paper.

If these individuals are removed, we can assign a gender to 75.2% of the remaining 439,642 authors.

28. M. F. Fox, "Gender, Science, and Academic Rank: Key Issues and Approaches," *Quantitative Science Studies* 1, no. 3 (2020): 1001–1006.

29. M. F. Fox and I. Nikivincze, "Being Highly Prolific in Academic Science: Character-istics of Individuals and Their Departments," *Higher Education* 81, no. 6 (2021); 1237–1255.

30. Jonathan R. Cole and Harriet Zuckerman, "The Productivity Puzzle: Persistence and Change in Patterns of Publication of Men and Women Scientists," *Advances in Motiva-tion and Achievement* 2 (1984): 217–258.

31. Gender assignment of Chinese authors, however, remains much more challenging, given the transliteration of Chinese names into roman characters. More details on the gender assignment algorithm can be found in the appendix.

32. Bogdan Denis Denitch, *The Legitimation of a Revolution: The Yugoslav Case* (New Haven, CT: Yale University Press, 1976).

33. Sabrina P. Ramet, ed., *Gender Politics in the Western Balkans: Women and Society in Yugoslavia and the Yugoslav Successor States* (University Park: Pennsylvania State University Press, 1999).

34. Qatar is an interesting case in terms of gender parity. Women represent only 25% of the population, due to the large number of male migrant laborers and the calculation of the population. The result is that Qatar has higher levels of gender parity in science and across the labor force. See World Bank, "Labor Force Participation Rate, Female (% of Female Population Ages 15+) (Modeled ILO Estimate)," International Labor Organization, ILO-STAT Database, January 29, 2021, https://data.worldbank.org/indicator/SL.TLF.CACT.FE .ZS?end=2019&start=1990&view=chart.

35. Calculated as those who have contributed to at least one paper over the 2008–2020 period. Rates of unique authors reported here are higher than what is provided by the United Nations Education, Scientific and Cultural Organization (UNESCO) on the percentage of re-searchers by gender, which the international organization establishes at 29.3% for 2016. The difference between the percentage of women researchers at UNESCO and in our dataset is likely due to the more encompassing definition of researchers used at UNESCO, which is based on surveys, and includes any personnel involved in research and development. Therefore, it includes applied and industrial researchers who do not necessarily have doctoral degrees and are less likely to publish, and it overrepresents male-dominated disciplines (such as engineer-ing). UNESCO, *Women in Science*, UIS Fact Sheet No. 55 (UNESCO Institute for Statistics, June 2019), http://uis.unesco.org/sites/default/files/documents/fs55-women-in-science-2019-en .pdf.

36. Several of the thirty countries have a small number of papers; 10,000 is used here as a threshold for highlighting the counties with the largest research communities.

37. World averages are calculated at the paper level and are therefore heavily reflective of the top-producing countries.

38. UNESCO, *Cracking the Code: Girls' and Women's Education in Science, Technology, Engineering, and Mathematics (STEM)* (Paris: UNESCO, 2017).

39. Mitra K. Shavarini, "The Feminisation of Iranian Higher Education," *International Review of Education* 51, no. 4 (July 2005): 329–347.

40. Zakiyyah Wahab, "Universities in Iran Put Limits on Women's Options," *New York Times*, August 20, 2012.

41. "Data," World Bank, accessed May 25, 2021, https://data.worldbank.org/.

42. See Valerii Tishkov, "Women in Russian Politics," *Economic and Political Weekly* 28, no. 51 (December 18, 1993): 2837–2840.

43. In economic terms, a substitution effect is when sales of a product decreases when consumers move to a less expensive alternative. In applying this to the scientific workforce, we see that when men are not present, women enter the workforce, but the result is a decrease in both economic and social capital. *Brain drain* typically refers to the loss of highly skilled workers due to mobility. In this case, it is a case of both mobility and morbidity.

44. OECD and European Observatory on Health Systems and Policies, *Latvia: Country Health Profile 2017, State of Health in the EU* (Paris: OECD; Brussels: European Observatory on Health Systems and Policies, 2017), https://doi.org/10.1787/9789264283466-en; Julia Smirnova and Weiyi Cai, "See Where Women Outnumber Men around the World (and Why)," *Washington Post,* August 19, 2015. At 52.2%, Latvia also has the second-highest percentage of women researchers according to UNESCO. See UNESCO, *Women in Science.*

45. Ukraine's percentage of women researchers (44.7% according to UNESCO) is also among the highest in Europe. See UNESCO, *Women in Science.*

46. This finding resonates with that of Stoet and Geary, who found that gender disparities in pursuit of STEM degrees increased with national gender equality. Our interpretation, however, differs. Gijsbert Stoet and David C. Geary, "The Gender-Equality Paradox in Science, Technology, Engineering, and Mathematics Education," *Psychological Science* 29, no. 4 (February 2018): 581–593.

47. Cassidy R. Sugimoto, Chaoqun Ni, and Vincent Larivière, "On the Relationship between Gender Disparities in Scholarly Communication and Country-Level Development Indicators," *Science and Public Policy* 42, no. 6 (December 2015): 789–810.

48. J. Scott Long and Mary Frank Fox, "Scientific Careers: Universalism and Particularism," *Annual Review of Sociology* 21 (August 1995): 68.

49. This paper suggests that the emigration of highly skilled women is higher when the country is poorer and suggests equality in those immigrating to OECD countries: Jean-Christophe Dumont, John P. Martin, and Gilles Spielvogel, "Women on the Move: The Neglected Gender Dimension of the Brain Drain" (IZA Discussion Paper No. 2920, Institute for the Study of Labor, Bonn, Germany, July 2007), https://www.oecd.org/els/mig/40232336.pdf.

50. To be discussed in full in Chapter 5.

51. Glenn E. Curtis, ed., *Poland: A Country Study* (Washington, DC: Government Printing Office for the Library of Congress, 1992).

52. Magdalena Moskalewicz, "Today's Women's Strike Has Its Roots in Poland—Where Women Have a Lot to Be Angry About," *Washington Post,* March 8, 2017.

53. Irena E. Kotowska, "Discrimination against Women in the Labor Market in Poland during the Transition to a Market Economy," *Social Politics: International Studies in Gender, State and Society* 2, no. 1 (Spring 1995): 76–90.

54. Agnieszka Bielecka, "Poland No Friend to Women," Human Rights Watch, December 3, 2017, https://www.hrw.org/news/2017/12/03/poland-no-friend-women#; Moskalewicz, "Today's Women's Strike."

55. Ausma Cimdina, *Enlargement, Gender, and Governance: Work Package 3 for Latvia,* EU Framework 5, Project No. HPSE-CT-2002-00115, n.d., http://www.qub.ac.uk/egg/Summaries /Latvia-WP3summ.doc.

56. Astrida Neimanis, *Gender and Human Development in Latvia* (Riga: United Nations Development Programme, 1999), http://pdc.ceu.hu/archive/00001549/01/Gender_EN[1] .pdf.

57. World Bank, "Data."

58. Neimanis, *Gender and Human Development.*

59. "Salary Comparisons," European University Institute, accessed May 25, 2021, https:// www.eui.eu/ProgrammesAndFellowships/AcademicCareersObservatory/CareerComparisons /SalaryComparisons.

60. Levanon, England, and Allison, "Occupational Feminization and Pay."

61. Jacqueline Heinen and Stéphane Portet, *Religion, Politics, and Gender Equality in Poland* (Geneva: United Nations Research Institute for Social Development, September 2009).

62. This hierarchy has, of course, shifted over time. In the University of Cambridge order of academic precedence in 1694 (far before the granting of doctoral degrees), theology was the loftiest degree, with medicine and law in the middle, and arts and philosophy at the bottom of the academic hierarchy. William Clark, *Academic Charisma and the Origins of the Research University* (Chicago: University of Chicago Press, 2006).

63. Catherine D'Ignazio and Lauren F. Klein, *Data Feminism* (Boston: MIT Press, 2020), 123.

2. Collaboration

1. Julie Des Jardins, *The Madame Curie Complex: The Hidden History of Women in Science* (New York: Feminist Press at the City University of New York, 2010), 57. Unlike in other chapters, we will refer to the scholars by their first names here, due to the confusion caused by the shared marital names.

2. Des Jardins, 68.

3. Des Jardins, 84.

4. Des Jardins, 58, 55, 53.

5. S. Lal, "Giving Children Security: Mamie Phipps Clark and the Racialization of Child Psychology," *American Psychologist* 57, no. 1 (2002): 20.

6. The Dolls Test was administered to 253 Black children, ages three to seven. The children were selected from a segregated school in Arkansas ($n=134$) and an integrated school in Massachusetts ($n=119$). Each child was presented with four different dolls: two with white skin and yellow hair, and two with brown skin and black hair. They were asked to state the race of each doll, and which doll they preferred. The results demonstrated distinct racial awareness (reinforcing Mamie Clark's previous research) and a preference toward the white doll with yellow hair. The children were also more likely to assign positive attributes to the white doll, demonstrating that the Black children had absorbed the connotation of their own racial identity with negative traits. Leila McNeill, "How a Psychologist's Work on Race Identity Helped Overturn School Segregation in 1950s America," *Smithsonian Magazine*, October 26, 2017.

7. "Profile: Mamie Phipps Clark," Psychology's Feminist Voices, accessed July 21, 2022, https://feministvoices.com/profiles/mamie-phipps-clark.

8. "A Revealing Experiment: Brown v. Board and 'the Doll Test,'" NAACP Legal Defense and Educational Fund, accessed July 21, 2022, https://www.naacpldf.org/ldf-celebrates -60th-anniversary-brown-v-board-education/significance-doll-test.

9. Gerald Markowitz and David Rosner, *Children, Race, and Power: Kenneth and Mamie Clark's Northside Center* (Charlottesville: University of Press of Virginia, 1996), 304.

10. Margaret W. Rossiter, *Women Scientists in America*, vol. 2, *Before Affirmative Action, 1940–1972* (Baltimore: Johns Hopkins University Press, 1998), 254.

11. S. Ranganathan, "She Was a Star," in *Lilavati's Daughters: The Women Scientists of India*, ed. Rohini Godbole and Ram Ramaswamy (repr., Bangalore: Indian Academy of Sciences, 2016), 27–30, https://archive.org/details/Ao560IASLeelavathisDaughterFullBook.

12. D. Balasubramanian, "Darshan Ranganathan—a Tribute," *Current Science* 81, no. 2 (2001): 217–218.

13. Ranganathan, "She Was a Star."

14. Balasubramanian, "Darshan Ranganathan," 217.

15. Balasubramanian, 218.

16. Balasubramanian, 217.

17. Ranganathan, "She Was a Star," 27.

18. M. Biagioli and P. Galison, introduction to *Scientific Authorship: Credit and Intellectual Property in Science,* ed. M. Biagioli and P. Galison (2003; New York: Routledge, 2014), 4.

19. M. Terrall, "The Uses of Anonymity in the Age of Reason," in Biagioli and Galison, *Scientific Authorship,* 91–112.

20. Emma Pierson, "Are Female Scientists Hiding?," FiveThirtyEight, August 5, 2014, https://fivethirtyeight.com/features/are-female-scientists-hiding. However, in a recent analysis, it was found that women were underrepresented among those with initials. This is due to the use of initials in the field of high-energy physics, where there are a disproportionate number of male authors.

21. J. P. Birnholtz, "What Does It Mean to Be an Author? The Intersection of Credit, Contribution, and Collaboration in Science," *Journal of the American Society for Information Science and Technology* 57, no. 13 (2006): 1758–1770.

22. S. Wuchty, B. F. Jones, and B. Uzzi, "The Increasing Dominance of Teams in Production of Knowledge," *Science* 316 (2007): 1036–1039; V. Larivière et al., "Team Size Matters: Collaboration and Scientific Impact since 1900," *Journal of the Association for Information Science and Technology* 66, no. 7 (2015): 1323–1332; B. Jones, "The Burden of Knowledge and the 'Death of the Renaissance Man': Is Innovation Getting Harder?," *Review of Economic Studies* 76 (2009): 283–317.

23. Larivière et al., "Team Size Matters."

24. Blaise Cronin, "Hyperauthorship: A Postmodern Perversion or Evidence of a Structural Shift in Scholarly Communication Practices?," *Journal of the American Society for Information Science and Technology* 52, no. 7 (2001): 558–569; D. Castelvecchi, "Physics Paper Sets Record with More than 5,000 Authors," *Nature,* May 15, 2015, https://www.nature.com/articles/nature.2015.17567.

25. ATLAS collaboration and CMS collaboration, "Combinations of Single-Top-Quark Production Cross-Section Measurements and |f LV Vtb| Determinations at √ s= 7 and 8 TeV with the ATLAS and CMS Experiments," *Journal of High Energy Physics* 2019, no. 5 (2019): 88.

26. D. L. Bhatt et al., "International Prevalence, Recognition, and Treatment of Cardiovascular Risk Factors in Outpatients with Atherothrombosis," *Journal of the American Medical Association* 295, no. 2 (2006): 180–189.

27. In 2019, the average number of authors per paper in nuclear and particle physics was more than 58.0. It was far ahead of other fields of physics, such as fluids and plasmas (12.3) and nuclear technology (11.7), and astronomy and astrophysics (9.5). In medical sciences, the field with the highest number of authors per paper was cardiovascular system (11.2).

28. C. R. Sugimoto and V. Larivière, *Measuring Research: What Everyone Needs to Know* (New York: Oxford University Press, 2018).

29. S. Shapin, "The Invisible Technician," *American Scientist* 77, no. 6 (1989): 554–563.

30. D. Pontille, "Scientific Authorship: Credit and Intellectual Property in Science," *Revue française de sociologie* 45, no. 2 (2004): 374–377; M. Biagioli, "Rights or Rewards," in Biagioli and Galison, *Scientific Authorship,* 253–280; J. P. Birnholtz, "What Does It Mean to Be an Author? The Intersection of Credit, Contribution, and Collaboration in Science," *Journal of the American Society for Information Science and Technology* 57, no. 13 (2006): 1758–1770.

31. A. Paul-Hus et al., "The Sum of It All: Revealing Collaboration Patterns by Combining Authorship and Acknowledgements," *Journal of Informetrics* 11, no. 1 (2017): 80–87.

32. H. A. Zuckerman, "Patterns of Name Ordering among Authors of Scientific Papers: A Study of Social Symbolism and Its Ambiguity," *American Journal of Sociology* 74, no. 3 (1968): 276–291.

33. L. Waltman, "An Empirical Analysis of the Use of Alphabetical Authorship in Scientific Publishing," *Journal of Informetrics* 6, no. 4 (2012): 700–711.

34. V. Larivière et al., "Contributorship and Division of Labor in Knowledge Production," *Social Studies of Science* 46, no. 3 (2016): 417–435.

35. Biagioli, "Rights or Rewards."

36. Biagioli, 269.

37. S. Sismondo, "Ghosts in the Machine: Publication Planning in the Medical Sciences," *Social Studies of Science* 39, no. 2 (2009): 171–198.

38. G. Mowatt et al., "Prevalence of Honorary and Ghost Authorship in Cochrane Reviews," *Journal of the American Medical Association* 287, no. 21 (2002): 2769–2771.

39. J. S. Wislar et al., "Honorary and Ghost Authorship in High Impact Biomedical Journals: A Cross Sectional Survey," *British Medical Journal* 343 (2011): d6128.

40. P. C. Gøtzsche et al., "Ghost Authorship in Industry-Initiated Randomised Trials," *PLoS Medicine* 4, no. 1 (2007): e19.

41. S. Jabbehdari and J. P. Walsh, "Authorship Norms and Project Structures in Science," *Science, Technology, and Human Values* 42, no. 5 (2017): 872–900; Shapin, "Invisible Technician."

42. C. Haeussler and H. Sauermann, "Credit Where Credit Is Due? The Impact of Project Contributions and Social Factors on Authorship and Inventorship," *Research Policy* 42, no. 3 (2013): 689.

43. A. Flanagin et al., "Prevalence of Articles with Honorary Authors and Ghost Authors in Peer-Reviewed Medical Journals," *Journal of the American Medical Association* 280, no. 3 (1998): 222–224.

44. B. Cronin, *The Scholar's Courtesy: The Role of Acknowledgement in the Primary Communication Process* (London: Taylor Graham, 1995).

45. Paul-Hus et al., "Sum of It All."

46. Jabbehdari and Walsh, "Authorship Norms."

47. A. Paul-Hus et al., "Who Are the Acknowledgees? An Analysis of Gender and Academic Status," *Quantitative Science Studies* 1, no. 2 (2020): 582–598.

48. "The New ICMJE Recommendations (August 2013)," International Committee of Medical Journal Editors, accessed September 15, 2021, http://www.icmje.org/news-and -editorials/new_rec_aug2013.html. This fourth criterion was added in 2013.

49. Z. Shapira, "'I've Got a Theory Paper—Do You?': Conceptual, Empirical, and Theoretical Contributions to Knowledge in the Organizational Sciences," *Organization Science* 22, no. 5 (2011): 1312–1321; P. J. Ågerfalk, "Insufficient Theoretical Contribution: A Conclusive Rationale for Rejection?," *European Journal of Information Systems* 23 (2014): 593–599; D. W. Aksnes, "Citation Rates and Perceptions of Scientific Contribution," *Journal of the American Society for Information Science and Technology* 57, no. 2 (2006): 169–185.

The Nobel Prize is also more likely to be awarded for theoretical rather than empirical contributions. See E. Garfield and A. Welljams-Dorof, "Of Nobel Class: A Citation Perspective on High Impact Research Authors," *Theoretical Medicine* 13, no. 2 (1992): 117–135; and Y. Gingras and M. L. Wallace, "Why It Has Become More Difficult to Predict Nobel Prize Winners: A Bibliometric Analysis of Nominees and Winners of the Chemistry and Physics Prizes (1901–2007)," *Scientometrics* 82, no. 2 (2010): 401–412.

50. G. Abramo, C. A. D'Angelo, and G. Murgia, "Gender Differences in Research Collaboration," *Journal of Informetrics* 7, no. 4 (2013): 811–822.

51. Larivière et al., "Team Size Matters"; V. Larivière, Y. Gingras, and É. Archambault, "Canadian Collaboration Networks: A Comparative Analysis of the Natural Sciences, Social Sciences, and the Humanities," *Scientometrics* 68, no. 3 (2006): 519–533.

52. This is particularly true in biomedical and natural sciences.

53. B. Macaluso et al., "Is Science Built on the Shoulders of Women? A Study of Gender Differences in Contributorship," *Academic Medicine* 91, no. 8 (2016): 1136–1142.

54. L. Holman and C. Morandin, "Researchers Collaborate with Same-Gendered Colleagues More Often than Expected across the Life Sciences," *PLOS ONE* 14, no. 4 (2019): e0216128M; Kwiek and W. Roszka, "Gender-Based Homophily in Research: A Large-Scale Study of Man-Woman Collaboration," *Journal of Informetrics* 15, no. 3 (2021): 101171.

55. N. Robinson-Garcia et al., "Meta-research: Task Specialization across Research Careers," *eLife* 9 (2020): e60586; S. Milojevic, F. Radicchi, and J. P. Walsh, "Changing Demographics of Scientific Careers: The Rise of the Temporary Workforce," *Proceedings of the National Academy of Sciences* 115, no. 50 (2018): 12616–12623.

56. It is at 39% for papers with two authors.

57. M. F. Fox and S. Mohapatra, "Social-Organizational Characteristics of Work and Publication Productivity among Academic Scientists in Doctoral-Granting Departments," *Journal of Higher Education* 78, no. 5 (2007): 542–571.

58. C. S. Wagner and L. Leydesdorff, "Network Structure, Self-Organization, and the Growth of International Collaboration in Science," *Research Policy* 34, no. 10 (2005): 1608–1618.

59. É Archambault et al., "Scale-Adjusted Metrics of Scientific Collaboration," in *Proceedings of ISSI 2011: The 13th International Conference of the International Society for Scientometrics and Informetrics, Durban, South Africa, July 4–7, 2011,* ed. Ed Noyons, Patrick Ngulube, and Jacqueline Leta (ISSI, Leiden University, and University of Zululand, 2011), 78–88; L. Waltman, R. J. Tijssen, and N. J. van Eck, "Globalisation of Science in Kilometres," *Journal of Informetrics* 5, no. 4 (2011): 574–582.

60. P. Bourdieu, *Science of Science and Reflexivity* (Cambridge, UK: Polity, 2004).

61. C. Cockburn, *Machinery of Dominance: Women, Men, and Technical Know-How* (Boston: Northeastern University Press, 1988); A. Witz, *Professions and Patriarchy* (New York: Routledge, 1992).

62. J. S. Long and M. F. Fox, "Scientific Careers: Universalism and Particularism," *Annual Review of Sociology* 21 (August 1995): 45–71; G. Derrick et al., "Models of Parenting and Its Effect on Academic Productivity: Preliminary Results from an International Survey," in *Proceedings of the 17th International Conference on Scientometrics and Informetrics* (Rome: ISSI, 2019), 1670–1676.

63. S. Sarabipour et al., "Changing Scientific Meetings for the Better," *Nature Human Behaviour* 5 (2021): 296–300; J. Wu et al., "Virtual Meetings Promise to Eliminate Geographical and Administrative Barriers and Increase Accessibility, Diversity and Inclusivity," *Nature Biotechnology* 40 (2022): 133–137.

64. K. R. Myers et al., "Unequal Effects of the COVID-19 Pandemic on Scientists," *Nature Human Behaviour* 4 (2020): 880–883.

65. M. Gibbons et al., *The New Production of Knowledge: The Dynamics of Science and Research in Contemporary Societies* (London: Sage, 1994).

66. V. Larivière et al., "Vanishing Industries and the Rising Monopoly of Universities in Published Research," *PLOS ONE* 13, no. 8 (2018): e0202120.

67. Gibbons et al., *New Production of Knowledge.*

68. M. W. Jackson, "Can Artisans Be Scientific Authors?," in Biagioli and Galison, *Scientific Authorship,* 113–132.

69. V. Tartari and A. Salter, "The Engagement Gap: Exploring Gender Differences in University–Industry Collaboration Activities," *Research Policy* 44, no. 6 (2015): 1176–1191.

70. C. R. Sugimoto et al., "The Academic Advantage: Gender Disparities in Patenting," *PLOS ONE* 10, no. 5 (2015): e0128000.

71. C. Ni et al., "The Gendered Nature of Authorship," *Science Advances* 7, no. 36 (2021): eabe4639.

72. These data are taken from an international survey of 5,575 respondents conducted in 2016. For more details, see Ni et al.

73. C. R. King et al., "Peer Review, Authorship, Ethics, and Conflict of Interest," *Image: Journal of Nursing Scholarship* 29, no. 2 (1997): 163–168.

74. King et al.

75. Ni et al., "Gendered Nature of Authorship."

76. H. Zuckerman, *Scientific Elite: Nobel Laureates in the United States* (New York: Free Press, 1977).

77. G. F. Gordukalova, review of *Scientific Elite: Nobel Laureates in the United States,* by Harriet Zuckerman, *Library Quarterly: Information, Community, Policy* 67, no. 3 (1997): 306–308.

78. R. K. Merton, "The Matthew Effect in Science, II: Cumulative Advantage and the Symbolism of Intellectual Property," *Isis* 79, no. 4 (1988): 607.

79. M. W. Rossiter, "The Matthew Matilda Effect in Science," *Social Studies of Science* 23, no. 2 (1993): 325–341.

80. Larivière et al., "Contributorship and Division of Labor."

81. A. Kiopa, J. Melkers, and Z. E. Tanyildiz, "Women in Academic Science: Mentors and Career Development," in *Women in Science and Technology,* ed. K. Prpić, L. Oliveira, and S. Hemlin (Zagreb: Institute for Social Research, 2009), 55.

82. B. R. Ragins and D. B. McFarlin, "Perceptions of Mentor Roles in Cross-Gender Mentoring Relationships," *Journal of Vocational Behavior* 37, no. 3 (1990): 321–339.

83. G. F. Dreher and J. A. Chargois, "Gender, Mentoring Experiences, and Salary Attainment among Graduates of an Historically Black University," *Journal of Vocational Behavior* 53, no. 3 (1998): 401–416; C. Hilmer and M. Hilmer, "Women Helping Women, Men Helping Women? Same-Gender Mentoring, Initial Job Placements, and Early Career Publishing Success for Economics PhDs," *American Economic Review* 97, no. 2 (2007): 422–426; G. F. Dreher and T. H. Cox Jr., "Race, Gender, and Opportunity: A Study of Compensation Attainment and the Establishment of Mentoring Relationships," *Journal of Applied Psychology* 81, no. 3 (1996): 297.

84. P. Gaule and M. Piacentini, "An Advisor like Me? Advisor Gender and Post-graduate Careers in Science," *Research Policy* 47, no. 4 (2018): 805–813.

85. M. Kogovšek and I. Ograjenšek, "Effects of the Same-Gender vs. Cross-Gender Mentoring on a Protégé Outcome in Academia: An Exploratory Study," *Advances in Methodology and Statistics* 16, no. 1 (2019): 61–78; J. B. Main, "Gender Homophily, Ph.D. Completion, and Time to Degree in the Humanities and Humanistic Social Sciences," *Review of Higher Education* 37, no. 3 (2014): 349–375; D. Neumark and R. Gardecki, "Women Helping Women? Role-Model and Mentoring Effects on Female Ph.D. Students in Economics," *Journal of Human Resources* 33, no. 1 (1996): 220–246.

86. D. Singh Chawla, "Assigning Authorship for Research Papers Can Be Tricky. These Approaches Can Help," *Science,* December 20, 2018, https://www.science.org/content /article/assigning-authorship-research-papers-can-be-tricky-these-approaches-can-help.

87. "Guidance on Authorship in Scholarly or Scientific Publications," Office of the Provost, Yale University, accessed July 22, 2022, https://provost.yale.edu/policies/academic-integrity /guidance-authorship-scholarly-or-scientific-publications.

88. M. Osterloh and B. S. Frey, "Ranking Games," *Evaluation Review* 39, no. 1 (2015): 102–129.

3. Contributorship

1. Margaret W. Rossiter, *Women Scientists in America,* vol. 1, *Struggles and Strategies to 1940* (Baltimore: Johns Hopkins University Press, 1982), 53.

2. Rossiter, 54.

3. Julie Des Jardins, *The Madame Curie Complex: The Hidden History of Women in Science* (New York: Feminist Press at the City University of New York, 2010), 97.

4. Rossiter, *Women Scientists in America,* 57.

5. Des Jardins, *Madame Curie Complex,* 99, 100.

6. Des Jardins, 90, 95–96.

7. Des Jardins, 89.

8. Des Jardins, 94, 96.

9. Margaret W. Rossiter, *Women Scientists in America,* vol. 3, *Forging a New World since 1972* (Baltimore: Johns Hopkins University Press, 2012), 308.

10. R. Ignotofsky, *Women in Science: 50 Fearless Pioneers Who Changed the World* (Berkeley: Ten Speed, 2016).

11. Radcliffe functioned as the women's parallel to the then all-men Harvard College. It was considered one of the Seven Sisters colleges, together with Bryn Mawr, Wellesley, Smith, Barnard, Mount Holyoke, and Vassar. It remained separate from Harvard until 1999, when it formally consolidated.

12. As reported by Payne-Gaposchkin in Des Jardins, *Madame Curie Complex,* 108.

13. Ignotofsky, *Women in Science,* 51.

14. M. Biagioli and P. Galison, eds., *Scientific Authorship: Credit and Intellectual Property in Science* (2003; New York: Routledge, 2014); D. Pontille, *La signature scientifique: Une sociologie pragmatique de l'attribution* (Paris: CNRS Éditions, 2004).

15. H. A. Zuckerman, "Patterns of Name Ordering among Authors of Scientific Papers: A Study of Social Symbolism and Its Ambiguity," *American Journal of Sociology* 74, no. 3 (1968): 276.

16. Zuckerman, 277.

17. See Chapter 2 for more on women's place on the byline in collaborative teams. For the rise of coauthorship, see V. Larivière et al., "Team Size Matters: Collaboration and Scientific Impact since 1900," *Journal of the Association for Information Science and Technology* 66, no. 7 (2015): 1323–1332.

18. D. Rennie, V. Yank, and L. Emanuel, "When Authorship Fails: A Proposal to Make Contributors Accountable," *Journal of the American Medical Association* 278, no. 7 (1997): 579, 580.

19. R. Smith, "Authorship: Time for a Paradigm Shift? The Authorship System Is Broken and May Need a Radical Solution," *British Medical Journal* 314 (1997): 992; R. Horton, "The Signature of Responsibility," *Lancet* 350 (1997): 5–6.

20. R. Smith, "Authorship Is Dying: Long Live Contributorship—The *BMJ* Will Publish Lists of Contributors and Guarantors to Original Articles," *British Medical Journal* 315, no. 7110 (1997): 696; Horton, "Signature of Responsibility."

21. See B. Macaluso, "Qui fait quoi? Analyse des libellés de contribution dans les articles savants" (MA thesis, Université de Montréal, 2015), 29–26, https://papyrus.bib.umontreal.ca/xmlui/handle/1866/12541.

22. V. Larivière et al., "Contributorship and Division of Labor in Knowledge Production," *Social Studies of Science* 46, no. 3 (2016): 417–435; B. Macaluso et al., "Is Science Built on the Shoulders of Women?," *Academic Medicine* 91, no. 8 (2016): 1136–1142.

23. Adele E. Clarke and Joan H. Fujimura, "1. What Tools? Which Jobs? Why Right?," in *The Right Tools for the Job: At Work in Twentieth-Century Life Sciences*, ed. Adele E. Clarke and Joan H. Fujimura (Princeton, NJ: Princeton University Press, 1992), 4; Andrew R. Pickering, "Knowledge, Practice, and Mere Construction," *Social Studies of Science* 20 (1990): 682–729.

24. Larivière et al., "Contributorship and Division of Labor."

25. These findings were replicated in subsequent analyses by H. Sauermann and C. Haeussler, "Authorship and Contribution Disclosures," *Science Advances* 3, no. 11 (2017): e1700404; and E. A. Corrêa Jr. et al., "Patterns of Authors Contribution in Scientific Manuscripts," *Journal of Informetrics* 11, no. 2 (2017): 498–510.

26. Macaluso et al., "Is Science Built?"

27. Macaluso et al.

28. This may reinforce what Duch observed: women tend to have fewer resources and are therefore less likely to engage in fields requiring high resources. It also seems to keep them from being able to gain authorship for this type of contribution. J. Duch et al., "The Possible Role of Resource Requirements and Academic Career-Choice Risk on Gender Differences in Publication Rate and Impact," *PLOS ONE* 7, no. 12 (2012): e51332.

29. Macaluso et al., "Is Science Built?"

30. H. Atkins, "Author Credit: PLOS and CRediT Update," *Official PLOS Blog,* July 8, 2016, http://blogs.plos.org/plos/2016/07/author-credit-plos-and-credit-update/.

31. I. Hames, *Report on the International Workshop on Contributorship and Scholarly Attribution* (Harvard University and the Wellcome Trust, May 16, 2012), https://projects .iq.harvard.edu/files/attribution_workshop/files/iwcsa_report_final_18sept12.pdf.

32. L. Allen et al., "Publishing: Credit Where Credit Is Due," *Nature* 508, no. 7496 (2014): 312–313.

33. The typology is now solely managed by the National Information Standards Organization. See "CRediT—Contributor Roles Taxonomy," NISO, accessed July 25, 2022, http:// credit.niso.org/.

34. M. K. McNutt et al., "Transparency in Authors' Contributions and Responsibilities to Promote Integrity in Scientific Publication," *Proceedings of the National Academy of Sciences* 115, no. 11 (2018): 2557–2560; L. Allen, A. O'Connell, and V. Kiermer, "How Can We Ensure Visibility and Diversity in Research Contributions? How the Contributor Role Taxonomy (CRediT) Is Helping the Shift from Authorship to Contributorship," *Learned Publishing* 32, no. 1 (2019): 71–74.

35. Macaluso et al., "Is Science Built?"

36. V. Larivière, D. Pontille, and C. R. Sugimoto, "Investigating the Division of Scientific Labor Using the Contributor Roles Taxonomy (CRediT)," *Quantitative Science Studies* 2, no. 1 (2021): 111–128.

37. As shown in Chapter 1, women account for 19% of authors in computer science, while it is more than two times higher in biology, biomedical sciences, and clinical medicine.

38. Larivière et al., "Contributorship and Division of Labor."

39. N. Robinson-Garcia et al., "Task Specialization and Its Effects on Research Careers," *eLife* 9 (2020): e60586.

40. Robinson-Garcia et al.

41. K. Siler, V. Larivière, and C. R. Sugimoto, "The Diverse Niches of Megajournals: Specialism within Generalism," *Journal of the Association for Information Science and Technology* 71, no. 7 (2020): 800–816.

42. We used the Science Citation Index and Social Science Citation Index as the sampling frame for this study. We excluded the Arts and Humanities Citation Index because of the low collaboration rates—and therefore division of labor—found in those disciplines. We selected only those papers published in 2016 with between three and ten authors. In total, this yielded a sample of 132,917 unique corresponding authors.

43. More details about the population, including response rates by discipline and country, can be found in the appendix.

44. Larivière et al., "Contributorship and Division of Labor"; Macaluso et al., "Is Science Built?"

45. J. D. Watson and F. H. C. Crick, "Molecular Structure of Nucleic Acids: A Structure for Deoxyribose Nucleic Acid," *Nature* 171 (1953): 737–738.

46. C. R. Sugimoto and V. Larivière, "Perspectives: Giving Credit Where It Is Due," *Chemical and Engineering News* 94, no. 35 (2016): 32–33.

47. Robinson-Garcia et al., "Task Specialization."

48. M. F. Fox and S. Mohapatra, "Social-Organizational Characteristics of Work and Publication Productivity among Academic Scientists in Doctoral-Granting Departments," *Journal of Higher Education* 78, no. 5 (2007): 542–571; G. Sonnert and G. J. Holton, *Who Succeeds in Science? The Gender Dimension* (New Brunswick, NJ: Rutgers University Press, 1995).

49. Introduced in Chapter 2.

50. F. C. Fang, J. W. Bennett, and A. Casadevall, "Males Are Overrepresented among Life Science Researchers Committing Scientific Misconduct," *mBio* 4, no. 1 (2013): e00640-12.

51. Larivière et al., "Contributorship and Division of Labor"; Sauermann and Haeussler, "Authorship and Contribution Disclosures"; Corrêa et al., "Patterns of Authors Contribution."

52. C. Ni et al., "The Gendered Nature of Authorship," *Science Advances* 7, no. 36 (2021): eabe4639.

53. S. R. Barley and B. A. Bechky, "In the Backrooms of Science: The Work of Technicians in Science Labs," *Work and Occupations* 21, no. 1 (1994): 85.

54. Barley and Bechky, 121.

55. J. P. Birnholtz, "What Does It Mean to Be an Author? The Intersection of Credit, Contribution, and Collaboration in Science," *Journal of the American Society for Information Science and Technology* 57, no. 13 (2006): 1758–1770.

56. Sauermann and Haeussler, "Authorship and Contribution Disclosures."

57. P. Wouters et al., "Rethinking Impact Factors: Better Ways to Judge a Journal," *Nature* 569 (2019): 621–623.

4. Funding

1. This story is drawn from Julie Des Jardins, *The Madame Curie Complex: The Hidden History of Women in Science* (New York: Feminist Press at the City University of New York, 2010).

2. Des Jardins, 25.

3. J. B. Sloan, "The Founding of the Naples Table Association for Promoting Scientific Research by Women, 1897," *Signs: Journal of Women in Culture and Society* 4, no. 1 (1978): 208–216.

4. Sloan.

5. Sloan, 210.

6. Sloan, 212–213.

7. R. A. Young, "On the Excretory Apparatus in Paramecium," *Science* 60, no. 1550 (1924): 244.

8. Marilyn Ogilvie and Joy Harvey, *The Biographical Dictionary of Women in Science: Pioneering Lives from Ancient Times to the Mid-20th Century* (New York: Routledge, 2000).

9. A. B. Jena et al., "Sex Differences in Academic Rank in US Medical Schools in 2014," *Journal of the American Medical Association* 314, no. 11 (2015): 1149–1158; C. Bloch, E. K. Graversen, and H. S. Pedersen, "Competitive Research Grants and Their Impact on Career Performance," *Minerva* 52, no. 1 (2014): 77–96.

10. E. J. Hackett, "Science as a Vocation in the 1990s: The Changing Organizational Culture of Academic Science," *Journal of Higher Education* 61, no. 3 (1990): 241–279.

11. P. Stephan, *How Economics Shapes Science* (Cambridge, MA: Harvard University Press, 2015).

12. E. L. Pier et al., "Low Agreement among Reviewers Evaluating the Same NIH Grant Applications," *Proceedings of the National Academy of Sciences* 115, no. 12 (2018): 2952–2957; J. Angermüller, "Beyond Excellence: An Essay on the Social Organization of the Social Sciences and Humanities," *Sociologica* 4, no. 3 (2010): 1–16; G. Mallard, M. Lamont, and J. Guetzkow, "Fairness as Appropriateness: Negotiating Epistemological Differences in Peer Review," *Science, Technology, and Human Values* 34, no. 5 (2009): 573–606.

13. H. Abdoul et al., "Peer Review of Grant Applications: Criteria Used and Qualitative Study of Reviewer Practices," *PLOS ONE* 7, no. 9 (2012): e46054; Pier et al., "Low Agreement among Reviewers"; E. L. Pier et al., "'Your Comments Are Meaner than Your Score': Score Calibration Talk Influences Intra- and Inter-panel Variability during Scientific Grant Peer Review," *Research Evaluation* 26, no. 1 (2017): 1–14.

14. M. Lamont, *How Professors Think: Inside the Curious World of Academic Judgment* (Cambridge, MA: Harvard University Press, 2009), 241.

15. G. Vallée-Tourangeau et al., "Peer Reviewers' Dilemmas: A Qualitative Exploration of Decisional Conflict in the Evaluation of Grant Applications in the Medical Humanities and Social Sciences," *Humanities and Social Sciences Communications* 9, no. 1 (2022): 9.

16. Angermüller, "Beyond Excellence."

17. R. Merton, *The Sociology of Science: Theoretical and Empirical Investigations* (Chicago: University of Chicago Press, 1973).

18. R. Jagsi et al., "Similarities and Differences in the Career Trajectories of Male and Female Career Development Award Recipients," *Academic Medicine* 86, no. 11 (2011): 1315–1421; L. Sigelman and F. P. Scioli, "Retreading Familiar Terrain—Bias, Peer Review, and the NSF Political Science Program," *PS: Political Science and Politics* 20, no. 1 (1987): 62–69; R. Jagsi et al., "Sex Differences in Attainment of Independent Funding by Career Development Awardees," *Annals of Internal Medicine* 151, no. 11 (2009): 804–811; "Racial Discrimination in Science," *Economist*, August 20, 2011; S. D. Hosek, *Is There a Gender Bias in Federal Grant Programs?* (Santa Monica, CA: RAND, 2004), http://www.rand.org/pubs/research_briefs/RB9147/index1.html.

19. M. Brouns, "The Gendered Nature of Assessment Procedures in Scientific Research Funding: The Dutch Case," *Higher Education in Europe* 25, no. 2 (2000): 193–199; L. A. Hechtman et al., "NIH Funding Longevity by Gender," *Proceedings of the National Academy of Sciences* 115, no. 31 (2018): 7943–7948; A. M. Joshi, T. M. Inouye, and J. A. Robinson, "How Does Agency Workforce Diversity Influence Federal R&D Funding of Minority and Women Technology Entrepreneurs? An Analysis of the SBIR and STTR Programs, 2001–2011," *Small Business Economics* 50, no. 3 (2018): 499–519; C. D. Zhou et al., "A Systematic Analysis of UK Cancer Research Funding by Gender of Primary Investigator," *BMJ Open* 8, no. 4 (2018): e018625.

20. P. J. Boyle et al., "Gender Balance: Women Are Funded More Fairly in Social Science," *Nature News* 525, no. 7568 (2015): 181–183.

21. J. Duch et al., "The Possible Role of Resource Requirements and Academic Career-Choice Risk on Gender Differences in Publication Rate and Impact," *PLOS ONE* 7, no. 12 (2012): e51332.

22. T. J. Ley and B. H. Hamilton, "The Gender Gap in NIH Grant Applications," *Science* 322, no. 5907 (2008): 1472–1474.

23. Hechtman et al., "NIH Funding Longevity."

24. A. Kaatz et al., "Analysis of National Institutes of Health R01 Application Critiques, Impact, and Criteria Scores: Does the Sex of the Principal Investigator Make a Difference?," *Academic Medicine* 91, no. 8 (2016): 1080–1088; D. K. Ginther et al., "Race, Ethnicity, and NIH Research Awards," *Science* 333, no. 6045 (2011): 1015–1019.

25. H. O. Witteman et al., "Are Gender Gaps Due to Evaluations of the Applicant or the Science? A Natural Experiment at a National Funding Agency," *Lancet* 393, no. 10171 (2019): 531–540.

26. H. W. Marsh et al., "Gender Effects in the Peer Reviews of Grant Proposals: A Comprehensive Meta-analysis Comparing Traditional and Multilevel Approaches," *Review of Educational Research* 79, no. 3 (2009): 1290–1326.

27. C. R. Sugimoto and V. Larivière, *Measuring Research: What Everyone Needs to Know* (New York: Oxford University Press, 2018).

28. J. Mervis, "Lawmakers Want to Know: Do U.S. Women Face Bias in Winning Federal Research Grants?," *Science,* April 4, 2015, https://www.science.org/content/article/lawmakers-want-know-do-us-women-face-bias-winning-federal-research-grants.

29. Sigelman and Scioli, "Retreading Familiar Terrain."

30. Ginther et al., "Race, Ethnicity, and NIH."

31. A. Paul-Hus, N. Desrochers, and R. Costas, "Characterization, Description, and Considerations for the Use of Funding Acknowledgement Data in Web of Science," *Scientometrics* 108, no. 1 (2016): 167–182.

32. J. Rigby, "Systematic Grant and Funding Body Acknowledgement Data for Publications: New Dimensions and New Controversies for Research Policy and Evaluation," *Research Evaluation* 20, no. 5 (2011): 365–375.

33. L. Tang, G. Hu, and W. Liu, "Funding Acknowledgment Analysis: Queries and Caveats," *Journal of the Association for Information Science and Technology* 68, no. 3 (2017): 790–794; V. Larivière and C. R. Sugimoto, "Do Authors Comply with Mandates for Open Access?," *Nature* 562 (2018): 483–486.

34. C. Flaherty, "Refusing to Be Measured," *Inside Higher Ed,* May 11, 2016, https://www.insidehighered.com/news/2016/05/11/rutgers-graduate-school-faculty-takes-stand-against-academic-analytics.

35. Despite differences in funding acknowledgments practices across countries, comparable results were obtained for a subset of countries analyzed.

36. Duch et al., "Possible Role of Resource Requirements."

37. C. Beaudry and V. Larivière, "Which Gender Gap? Factors Affecting Researchers' Scientific Impact in Science and Medicine," *Research Policy* 45, no. 9 (2016): 1790–1817.

38. D. Murray, V. Lariviere, and C. R. Sugimoto, "A Balanced Portfolio? The Relationship between Gender and Funding for US Academic Professors" (paper presented at the 2017 Science, Technology, and Innovation Indicators conference, Paris, France).

39. Academic Analytics is a US consultancy that provides research data on papers published, funding, and awards obtained by individual researchers affiliated with about 400 academic institutions (at the time we obtained our data dump). It collects data from publicly available databases of research funding in the United States, such as the NSF, the NIH, and

other agencies (NOAA, NASA) and private foundations. It does not provide detailed information on awards but, rather, provides total amounts received as well as numbers of grants as principal investigator over the previous five years.

40. G. Ghiasi, V. Larivière, and C. R. Sugimoto, "On the Compliance of Women Engineers with a Gendered Scientific System," *PLOS ONE* 10, no. 12 (2015): e0145931.

41. Y. Gingras et al., "The Effects of Aging on Researchers' Publication and Citation Patterns," *PLOS ONE* 3, no. 12 (2008): e4048; K. R. Matthews et al., "The Aging of Biomedical Research in the United States," *PLOS ONE* 6, no. 12 (2011): e29738; G. Abramo, C. A. D'Angelo, and G. Murgia, "The Combined Effects of Age and Seniority on Research Performance of Full Professors," *Science and Public Policy* 43, no. 3 (2016): 301–319.

42. M. Lauer, "Long-Term Trends in the Age of Principal Investigators Supported for the First Time on NIH R01-Equivalent Awards," Extramural Nexus, November 18, 2021, https://nexus.od.nih.gov/all/2021/11/18/long-term-trends-in-the-age-of-principal-investigators-supported-for-the-first-time-on-nih-r01-awards/.

43. J. C. Williams, K. W. Phillips, and E. V. Hall, "Tools for Change: Boosting the Retention of Women in the STEM Pipeline," *Journal of Research in Gender Studies* 6, no. 1 (2016): 11–75.

44. The platform of the Liberal Party for the 2015 election made abundant references to issues related to women and gender and made strong commitment in this direction. Liberal Party of Canada, *Real Change: A New Plan for a Strong Middle Class* (2015), https://s3.documentcloud.org/documents/2448348/new-plan-for-a-strong-middle-class.pdf. Among others, the Canadian cabinet (that is, elected members of Parliament who hold ministerial positions) that emerged from the 2015 election had gender parity—a first in Canadian history.

45. "Equity, Diversity and Inclusion Action Plan," Canada Research Chairs, accessed December 24, 2021, https://www.chairs-chaires.gc.ca/program-programme/equity-equite/action_plan-plan_action-eng.aspx.

46. Natural Sciences and Engineering Research Council of Canada, *Guide for Applicants: Considering Equity, Diversity and Inclusion in Your Application*, 2017 ed., https://www.nserc-crsng.gc.ca/_doc/EDI/Guide_for_Applicants_EN.pdf.

47. Witteman et al., "Are Gender Gaps Due?"; H. O. Witteman, J. Haverfield, and C. Tannenbaum, "COVID-19 Gender Policy Changes Support Female Scientists and Improve Research Quality," *Proceedings of the National Academy of Sciences* 118, no. 6 (2021): e2023476118.

48. "Gender Equity Data Analysis—CIHR Competition Success Rates by Gender (Training Awards)," Canadian Institutes of Health Research, accessed December 24, 2021, http://www.cihr-irsc.gc.ca/e/50073.html; "Gender Equity Data Analysis—CIHR Competition Success Rates by Gender (Project Grant)," Canadian Institutes of Health Research, accessed December 24, 2021, http://www.cihr-irsc.gc.ca/e/50235.html.

49. Liisa Galea, "Gender Equity in CIHR Research Funding: Implications for Women's Health Research," Women's Health Research Institute, Our News, November 20, 2017, http://whri.org/gender-equity-in-cihr-research-funding-implications-for-womens-health-research/.

50. "Gender Equity Data Analysis—CIHR Competition Success Rates by Gender (All CIHR Grant Programs)," Canadian Institutes of Health Research, accessed December 24, 2021, http://www.cihr-irsc.gc.ca/e/50464.html.

51. "Equity, Diversity and Inclusion in CIHR Funding Programs," Canadian Institutes of Health Research, accessed December 24, 2021, https://cihr-irsc.gc.ca/e/52552.html.

52. An analogous—albeit more positive—phenomenon is observed for men in feminized professions. See J. S. Dill, K. Price-Glynn, and C. Rakovski, "Does the 'Glass Escalator'

Compensate for the Devaluation of Care Work Occupations? The Careers of Men in Low- and Middle-Skill Health Care Jobs," *Gender and Society* 30, no. 2 (2016): 334–360.

53. P. J. Boyle et al., "Gender Balance: Women Are Funded More Fairly in Social Science," *Nature* 525, no. 7568 (2015): 181–183.

54. V. Larivière et al., "Sex Differences in Research Funding, Productivity and Impact: An Analysis of Québec University Professors," *Scientometrics* 87, no. 3 (2011): 483–498.

55. See the promotions and tenure guidelines for the University of Southern California: University of Southern California, *University Committee on Appointments, Promotions, and Tenure, UCAPT Manual 2017* (Los Angeles: University of Southern California, 2017), https://policy.usc.edu/wp-content/uploads/2021/04/University-Committee-on-Appointments -Promotions-and-Tenure-Manual.pdf.

56. M. J. Lerchenmueller and O. Sorenson, "The Gender Gap in Early Career Transitions in the Life Sciences," *Research Policy* 47, no. 6 (2018): 1007–1017.

57. Hechtman et al., "NIH Funding Longevity."

58. C. R. Sugimoto et al., "Factors Affecting Sex-Related Reporting in Medical Research: A Cross-Disciplinary Bibliometric Analysis," *Lancet* 393, no. 10171 (2019): 550–559.

59. J. R. Pohlhaus et al., "Sex Differences in Application, Success, and Funding Rates for NIH Extramural Programs," *Academic Medicine* 86, no. 6 (2011): 759.

60. B. K. Swenor, B. Munoz, and L. M. Meeks, "A Decade of Decline: Grant Funding for Researchers with Disabilities 2008 to 2018," *PLOS ONE* 15, no. 3 (2020): e0228686; W. P. Wahls, "Biases in Grant Proposal Success Rates, Funding Rates and Award Sizes Affect the Geographical Distribution of Funding for Biomedical Research," *PeerJ* 4 (2016): e1917.

61. L. Bornmann, R. Mutz, and H. D. Daniel, "Gender Differences in Grant Peer Review: A Meta-analysis," *Journal of Informetrics* 1, no. 3 (2007): 226–238; R. Van der Lee and N. Ellemers, "Gender Contributes to Personal Research Funding Success in the Netherlands," *Proceedings of the National Academy of Sciences* 112, no. 40 (2015): 12349–12353; Witteman et al., "Are Gender Gaps Due?"

62. T. Bol, M. de Vaan, and A. van de Rijt, "Gender-Equal Funding Rates Conceal Unequal Evaluations," *Research Policy* 51, no. 1 (2022): 104399.

63. Joshi, Inouye, and Robinson, "How Does Agency Workforce?"

64. D. Murray et al., "Author-Reviewer Homophily in Peer Review," BioRxiv, 400515 (2019).

65. Nicola Jones, "Biochemist Chosen as Canada's Chief Science Adviser," *Nature*, September 26, 2017; "Office of the Chief Science Advisor," Government of Canada, accessed December 24, 2021, https://www.ic.gc.ca/eic/site/063.nsf/eng/h_97646.html.

66. "Athena Swan Charter," Advance HE, accessed December 24, 2021, https://www .advance-he.ac.uk/equality-charters/athena-swan-charter.

67. P. V. Ovseiko et al., "Effect of Athena SWAN Funding Incentives on Women's Research Leadership," *British Medical Journal* 2020, no. 371 (2020): m3975.

68. S. V. Rosser et al., "Athena SWAN and ADVANCE: Effectiveness and Lessons Learned," *Lancet* 393, no. 10171 (2019): 604.

69. Rosser et al.

70. J. DeAro, S. Bird, and S. M. Ryan, "NSF ADVANCE and Gender Equity: Past, Present and Future of Systemic Institutional Transformation Strategies," *Equality, Diversity and Inclusion* 38, no. 2 (2019): 131–139.

71. These institutions are the University of California, Davis; Boston University; the University of Massachusetts, Lowell; Arizona State University; and the University of California, Irvine.

72. "Equity, Diversity and Inclusion Action Plan."

73. "Results of Formal Review of Institutional Equity, Diversity, and Inclusion Action Plans," Canada Research Chairs, accessed December 24, 2021, https://www.chairs-chaires .gc.ca/program-programme/equity-equite/results_of_formal_review-resultats_de_l_evaluation _officielle-eng.aspx.

74. K. A. Smith et al., "Seven Actionable Strategies for Advancing Women in Science, Engineering, and Medicine," *Cell Stem Cell* 16, no. 3 (2015): 222.

75. C. R. Sugimoto and V. Larivière, "Indicators for Social Good," Centre for Science and Technology Studies, blog archive, May 15, 2019, https://www.cwts.nl/blog?article=n-r2w2c4.

76. "CWTS Leiden Ranking 2021," Center for Science and Technology Studies, accessed December 24, 2021, https://www.leidenranking.com.

77. G. C. Black and P. E. Stephan, "The Economics of University Science and the Role of Foreign Graduate Students and Postdoctoral Scholars," in *American Universities in a Global Market,* ed. C. T. Clotfelter (Chicago: University of Chicago Press, 2010), 129–161; V. Larivière, "On the Shoulders of Students? The Contribution of PhD Students to the Advancement of Knowledge," *Scientometrics* 90, no. 2 (2012): 463–481.

78. Stephan, *How Economics Shapes Science.*

5. Mobility

1. S. Spillman, "Institutional Limits: Christine Ladd-Franklin, Fellowships, and American Women's Academic Careers, 1880–1920," *History of Education Quarterly* 52, no. 2 (2012): 196–221.

2. Alice Hamilton, "Edith and Alice Hamilton: Students in Germany," *Atlantic Monthly,* March 1965, 131.

3. Margaret W. Rossiter, *Women Scientists in America,* vol. 1, *Struggles and Strategies to 1940* (Baltimore: Johns Hopkins University Press, 1982), 38.

4. Rossiter, 39.

5. Spillman, "Institutional Limits."

6. Rossiter, 43.

7. K. Zippel, "Pathways for Women in Global Science," in *Pathways, Potholes, and the Persistence of Women in Science,* ed. Enobong Hannah Branch (Lanham, MD: Lexington Books, 2016), 169–182; J. Lerner and R. Roy, "Numbers, Origins, Economic Value and Quality of Technically Trained Immigrants into the United States," *Scientometrics* 6, no. 4 (1984): 243–259; J. Vilcek and B. N. Cronstein, "A Prize for the Foreign-Born," *FASEB Journal* 20, no. 9 (2006): 1281–1283.

8. G. Laudel, "Migration Currents among the Scientific Elite," *Minerva* 43, no. 4 (2005): 377–395.

9. L. Ackers, "Internationalisation and Equality: The Contribution of Short-Stay Mobility to Progression in Science Careers," *Recherches Sociologiques et Anthropologiques* 41, no. 1 (2010): 83–103.

10. F. Huang, "Policy and Practice of the Internationalization of Higher Education in China," *Journal of Studies in International Education* 7, no. 3 (2003): 225–240. See also "Nearly 90% of All Chinese Students Return Home after Studying Abroad: MOE," *Global Times,* December 15, 2020, https://www.globaltimes.cn/content/1210043.shtml. These have also, historically, served to build capacity in the country of origin of those researchers. See R. Gagnon and D. Goulet, "Les 'boursiers d'Europe,' 1920–1959: La formation d'une élite scientifique au Québec," *Bulletin d'histoire politique* 20, no. 1 (2011): 60–71. The impact of these programs is mixed, with some studies suggesting that returnees outperform those who

stayed, whereas other studies demonstrate no higher performance for returnees. X. Lu and W. Zhang, "The Reversed Brain Drain: A Mixed-Method Study of the Reversed Migration of Chinese Overseas Scientists," *Science, Technology and Society* 20, no. 3 (2015): 279–299; J. Gibson and D. McKenzie, "Scientific Mobility and Knowledge Networks in High Emigration Countries: Evidence from the Pacific," *Research Policy* 43, no. 9 (2014): 1486–1495.

11. M. Schaer, J. Dahinden, and A. Toader, "Transnational Mobility among Early-Career Academics: Gendered Aspects of Negotiations and Arrangements within Heterosexual Couples," *Journal of Ethnic and Migration Studies* 43, no. 8 (2017): 1292.

12. L. Ackers, "Internationalisation, Mobility and Metrics: A New Form of Indirect Discrimination," *Minerva* 46, no. 4 (2008): 418.

13. R. B. Freeman, "Globalization of Scientific and Engineering Talent: International Mobility of Students, Workers, and Ideas and the World Economy," *Economics of Innovation and New Technology* 19, no. 5 (2010): 393–406.

14. S. Cohen et al., "Gender Discourses in Academic Mobility," *Gender, Work and Organization* 27, no. 2 (2020): 149–165.

15. H. Jöns, "Transnational Academic Mobility and Gender," *Globalisation, Societies and Education* 9, no. 2 (2011): 183–209; N. Bonney and J. Love, "Gender and Migration: Geographical Mobility and the Wife's Sacrifice," *Sociological Review* 39, no. 2 (1991): 335–348.

16. C. Cañibano, M. F. Fox, and F. J. Otamendi, "Gender and Patterns of Temporary Mobility among Researchers," *Science and Public Policy* 43, no. 3 (2016): 320–331; C. Cañibano et al., "Scientific Careers and the Mobility of European Researchers: An Analysis of International Mobility by Career Stage," *Higher Education* 80, no. 6 (2020): 1175–1193. It is interesting to note that this pattern may be inverted in certain regions. For example, in a study of African scientists, reverse trends were found (mobility differences were reduced by age). H. Prozesky and C. Beaudry, "Mobility, Gender and Career Development in Higher Education: Results of a Multi-country Survey of African Academic Scientists," *Social Sciences* 8, no. 6 (2019): 188.

17. H. Jons, "Transnational Mobility and the Spaces of Knowledge Production: A Comparison of Global Patterns, Motivations and Collaborations in Different Academic Fields," *Social Geography* 2 (2007): 97–114.

18. Cañibano, Fox, and Otamendi, "Gender and Patterns."

19. K. Jonkers, "Mobility, Productivity, Gender and Career Development of Argentinean Life Scientists," *Research Evaluation* 20, no. 5 (2011): 411–421.

20. Organisation for Economic Cooperation and Development, *The Global Competition for Talent: Mobility of the Highly Skilled* (Paris: Organisation for Economic Cooperation and Development, 2008); United Nations Education, Scientific and Cultural Organization, *UNESCO Science Report: The Race against Time for Smarter Development* (Paris: United Nations Education, Scientific and Cultural Organization, 2021); M. Åkerblom, "Constructing Internationally Comparable Indicators on the Mobility of Highly Qualified Workers," *STI Review* 27 (2002): 49–75.

21. M. Beine, R. Noël, and L. Ragot, "Determinants of the International Mobility of Students," *Economics of Education Review* 41 (2014): 40–54.

22. G. Laudel, "Studying the Brain Drain: Can Bibliometric Methods Help?," *Scientometrics* 57, no. 2 (2003): 215–237.

23. M. D. Heusse and G. Cabanac, "ORCID Growth and Field-Wise Dynamics of Adoption: A Case Study of the Toulouse Scientific Area," *Learned Publishing* (forthcoming, 2022); N. R. Smalheiser and V. I. Torvik, "Author Name Disambiguation," *Annual Review of Information Science and Technology* 43, no. 1 (2009): 1–43.

24. E. Caron and N. J. van Eck, "Large Scale Author Name Disambiguation Using Rule-Based Scoring and Clustering," in *Proceedings of the 19th International Conference on Sci-*

ence and Technology Indicators (Leiden: Center for Science and Technology Studies, Leiden University, 2014), 79–86.

25. R. Veugelers and L. Van Bouwel, "The Effects of International Mobility on European Researchers: Comparing Intra-EU and US Mobility," *Research in Higher Education* 56 (2015): 360–377; U. Sandström, "Combining Curriculum Vitae and Bibliometric Analysis: Mobility, Gender and Research Performance," *Research Evaluation* 18, no. 2 (2009): 135–142; C. R. Sugimoto et al., "Scientists Have Most Impact When They're Free to Move," *Nature* 550 (2017): 29–31; A. M. Petersen, "Multiscale Impact of Researcher Mobility," *Journal of the Royal Society Interface* 15, no. 146 (2018): 20180580; D. W. Aksnes et al., "Are Mobile Researchers More Productive and Cited than Non-mobile Researchers? A Large-Scale Study of Norwegian Scientists," *Research Evaluation* 22, no. 4 (2013): 215–223. For example, G. Halevi and colleagues and C. Cañibano and colleagues find no relationship between international mobility and publication productivity. G. Halevi, H. F. Moed, and J. Bar-Ilan, "Researchers' Mobility, Productivity and Impact: Case of Top Producing Authors in Seven Disciplines," *Publishing Research Quarterly* 32, no. 1 (2016): 22–37; C. Cañibano, J. Otamendi, and I. Andújar, "Measuring and Assessing Researcher Mobility from CV Analysis: The Case of the Ramón y Cajal Programme in Spain," *Research Evaluation* 17, no. 1 (2008): 17–31.

26. As in other chapters, this analysis relies on bibliometric data from Web of Science, for which we have disambiguated authors' names based on their given names and family names (for countries where family names are gendered). More details can be found in the appendix.

27. M. Lauer, "Long-Term Trends in the Age of Principal Investigators Supported for the First Time on NIH R01-Equivalent Awards," Extramural Nexus, November 18, 2021, https://nexus.od.nih.gov/all/2021/11/18/long-term-trends-in-the-age-of-principal-investigators-supported-for-the-first-time-on-nih-r01-awards/.

28. N. Robinson-Garcia et al., "The Many Faces of Mobility: Using Bibliometric Data to Measure the Movement of Scientists," *Journal of Informetrics* 13, no. 1 (2019): 50–63.

29. One might notice differences between the proportion of travelers and migrants noted here and in our previous work. This is largely due to the more expansive time window utilized in this book, which provides more time for migration. Sugimoto et al., "Scientists Have Most Impact."

30. Organisation for Economic Cooperation and Development, *Global Competition for Talent.*

31. This is similar to our findings in funding, where women were closer to parity in disciplines where funding was essential for research.

32. However, mobility was a modus operandi for hyperproductive mathematician Paul Erdős.

33. J. El-Ouahi, N. Robinson-García, and R. Costas, "Analyzing Scientific Mobility and Collaboration in the Middle East and North Africa," *Quantitative Science Studies* 2, no. 3 (2021): 1023–1047.

34. Sugimoto et al., "Scientists Have Most Impact."

35. Robinson-Garcia et al., "Many Faces of Mobility."

36. Sugimoto et al., "Scientists Have Most Impact."

37. Zippel, "Pathways for Women," 75.

38. A. Clauset, S. Arbesman, and D. B. Larremore, "Systematic Inequality and Hierarchy in Faculty Hiring Networks," *Science Advances* 1, no. 1 (2015): e1400005.

39. C. Cockburn, *Machinery of Dominance: Women, Men, and Technical Know-How* (Boston: Northeastern University Press, 1988); A. Witz, *Professions and Patriarchy* (New York: Routledge, 1992).

40. V. Larivière and R. Costas, "How Many Is Too Many? On the Relationship between Research Productivity and Impact," *PLOS ONE* 11, no. 9 (2016): e0162709.

41. Zippel, "Pathways for Women," 75.

42. Zippel.

43. "Arab States/North Africa," UN Women, accessed December 26, 2021, https://www.unwomen.org/en/where-we-are/arab-states-north-africa.

44. Zippel, "Pathways for Women."

45. Zippel; L. Ackers, "Managing Relationships in Peripatetic Careers: Scientific Mobility in the European Union," *Women's Studies International Forum* 27, no. 3 (2004): 189–201.

46. C. Woolston, "Postdoc Survey Reveals Disenchantment with Working Life," *Nature,* November 18, 2020.

47. For a review of issues surrounding mobility in science, see A. Geuna, ed., *Global Mobility of Research Scientists: The Economics of Who Goes Where and Why* (London: Academic Press, 2015), esp. chap. 7.

48. Sugimoto et al., "Scientists Have Most Impact."

49. The COVID-19 pandemic may provide a natural experiment to study mobility, as early studies on collaboration suggest: C. V. Fry et al., "Consolidation in a Crisis: Patterns of International Collaboration in Early COVID-19 Research," *PLOS ONE* 15, no. 7 (2020): e0236307.

50. Ackers, "Managing Relationships"; S. Kulis and D. Sicotte, "Women Scientists in Academia: Geographically Constrained to Big Cities, College Clusters, or the Coasts?," *Research in Higher Education* 43, no. 1 (2002): 1–30.

51. Ackers, "Managing Relationships."

52. Zippel, "Pathways for Women."

53. Zippel; K. A. Shauman and Y. Xie, "Geographic Mobility of Scientists: Sex Differences and Family Constraints," *Demography* 33, no. 4 (1996): 455–468.

54. Jöns, "Transnational Academic Mobility."

55. M. F. Fox, C. Fonseca, and J. Bao, "Work and Family Conflict in Academic Science: Patterns and Predictors among Women and Men in Research Universities," *Social Studies of Science* 41, no. 5 (2011): 715–735.

56. B. Keating, *Losing the Nobel Prize: A Story of Cosmology, Ambition, and the Perils of Science's Highest Honor* (New York: W. W. Norton, 2018), 152.

57. Meredith Wadman, "Disturbing Allegations of Sexual Harassment in Antarctica Leveled at Noted Scientist," *Science,* October 6, 2017, https://www.science.org/content/article/disturbing-allegations-sexual-harassment-antarctica-leveled-noted-scientist; Meredith Wadman, "Boston University Fires Geologist Found to Have Harassed Women in Antarctica," *Science,* April 12, 2019, https://www.science.org/content/article/boston-university-fires-geologist-who-sexually-harassed-women-antarctica.

58. National Academies of Sciences, Engineering, and Medicine, *Sexual Harassment of Women: Climate, Culture, and Consequences in Academic Sciences, Engineering, and Medicine* (Washington, DC: National Academies Press, 2018).

59. K. B. H. Clancy et al., "Survey of Academic Field Experiences (SAFE): Trainees Report Harassment and Assault," *PLOS ONE* 9, no. 7 (2014): e102172.

60. Jöns, "Transnational Academic Mobility."

6. Scientific Impact

1. There are six instances of dual prizes, of which two were organizations.

2. Margaret W. Rossiter, *Women Scientists in America,* vol. 1, *Struggles and Strategies to 1940* (Baltimore: Johns Hopkins University Press, 1982), 127.

3. Julie Des Jardins, *The Madame Curie Complex: The Hidden History of Women in Science* (New York: Feminist Press at the City University of New York, 2010), 5.

4. R. L. Sime, *From Exceptional Prominence to Prominent Exception: Lise Meitner at the Kaiser Wilhelm Institute for Chemistry,* Ergebnisse 24 (Berlin: Forschungsprogramm Geschichte der Kaiser-Wilhelm-Gesellschaft im Nationalsozialismus, 2005), https://www.mpiwg-berlin.mpg.de/KWG/Ergebnisse/Ergebnisse24.pdf.

5. Des Jardins, *Madame Curie Complex,* 43.

6. Sime, *From Exceptional Prominence.*

7. B. R. Brown, *Planck: Driven by Vision, Broken by War* (New York: Oxford University Press, 2015), 20.

8. Brown, 22.

9. Sime, *From Exceptional Prominence,* 151–152.

10. Sime, 151–152.

11. A *Mitarbeiterin* is more of a technical assistant than a lead investigator. Des Jardins, *Madame Curie Complex,* 161; Sime, *From Exceptional Prominence.*

12. Elisabeth Crawford and Ruth Lewin Sime, "A Nobel Tale of Postwar Injustice," *Physics Today* 50, no. 9 (1997): 32.

13. Wu's response: "Chinese people think that calling me the 'Chinese Madame Curie' is an endorsement and honor, but I do not quite feel that way." T. C. Chian, *Madame Wu Chien-Shiung: The First Lady of Physics Research* (Singapore: World Scientific, 2014), 177.

14. Chian, 173.

15. Given her work on nuclear fission, Meitner was invited to work on the Manhattan Project at Los Alamos. She rejected the offer, however, not wanting to have anything to do with a bomb.

16. Des Jardins, *Madame Curie Complex,* 159; Rossiter, *Women Scientists in America.*

17. "Chien-Shung Wu (1912–1997)," National Science Foundation, accessed December 26, 2021, https://www.nsf.gov/news/special_reports/medalofscience50/wu.jsp.

18. Chian, *Madame Wu Chien-Shiung.*

19. J. McBride, "Nobel Laureate Donna Strickland: 'I See Myself as a Scientist, Not a Woman in Science,'" *Guardian,* October 20, 2018.

20. D. Bazely, "Why Nobel Winner Donna Strickland Didn't Have a Wikipedia Page," *Washington Post,* October 8, 2018.

21. "Alfred Nobel's Will," Nobel Prize, accessed July 26, 2022, https://www.nobelprize.org/alfred-nobel/alfred-nobels-will/.

22. Including the Sveriges Riksbank Prize in Economic Sciences in Memory of Alfred Nobel, often referred to as the Nobel Prize in Economic Sciences.

23. Four in physics, seven in chemistry—with Marie Curie winning both, therefore counting as one individual in the total—twelve in physiology or medicine, and two in economics.

24. E. Garfield, "'Science Citation Index'—a New Dimension in Indexing," *Science* 144, no. 3619 (1964): 649–654.

25. F. Narin, *Evaluative Bibliometrics: The Use of Publication and Citation Analysis in the Evaluation of Scientific Activity* (Cherry Hill, NJ: Computer Horizons, 1976), 334–337.

26. C. R. Sugimoto and V. Larivière, *Measuring Research: What Everyone Needs to Know* (Oxford: Oxford University Press, 2018). Citation counts were also shown to favor Nobel laureates, as Zuckerman demonstrated in *Scientific Elite*: H. Zuckerman, *Scientific Elite: Nobel Laureates in the United States* (New York: Free Press, 1977).

27. E. Garfield, "Citation Indexing for Studying Science," *Nature* 227, no. 5259 (1970): 669–671.

28. K. Debackere and W. Glänzel, "Using a Bibliometric Approach to Support Research Policy Making: The Case of the Flemish BOF-Key," *Scientometrics* 59, no. 2 (2004): 253–276;

D. Hicks, "Performance-Based University Research Funding Systems," *Research Policy* 41, no. 2 (2012): 251–261; W. Quan, B. Chen, and F. Shu, "Publish or Impoverish: An Investigation of the Monetary Reward System of Science in China (1999–2016)," *Aslib Journal of Information Management* 69, no. 5 (2017): 486–502; J. P. Alperin et al., "How Significant Are the Public Dimensions of Faculty Work in Review, Promotion, and Tenure Documents?," *Humanities Commons,* preprint, 2018, https://hcommons.org/deposits/download/hc:21016/CONTENT/publicness-in-rpt-preprint.pdf/.

29. L. Bornmann and H. D. Daniel, "What Do Citation Counts Measure? A Review of Studies on Citing Behavior," *Journal of Documentation* 64, no. 1 (2008): 45–80.

30. C. R. Sugimoto, ed., *Theories of Informetrics and Scholarly Communication* (Berlin: De Gruyter, 2016), 426.

31. R. K. Merton, "The Matthew Effect in Science, II: Cumulative Advantage and the Symbolism of Intellectual Property," *Isis* 79, no. 4 (1988): 622.

32. G. Nigel Gilbert, "Referencing as Persuasion," *Social Studies of Science* 7, no. 1 (1977): 113–122.

33. W. S. Lamers et al., "Investigating Disagreement in the Scientific Literature," arXiv preprint, arXiv:2107.14641 (2021).

34. V. Larivière and Y. Gingras, "The Impact Factor's Matthew Effect: A Natural Experiment in Bibliometrics," *Journal of the American Society for Information Science and Technology* 61, no. 2 (2010): 424–427.

35. Bornmann and Daniel, "What Do Citation Counts Measure?"; L. Bornmann et al., "What Factors Determine Citation Counts of Publications in Chemistry Besides Their Quality?," *Journal of Informetrics* 6, no. 1 (2012): 11–18; N. Onodera and F. Yoshikane, "Factors Affecting Citation Rates of Research Articles," *Journal of the Association for Information Science and Technology* 66, no. 4 (2015): 739–764.

36. V. Larivière, K. Gong, and C. R. Sugimoto, "Citations Strength Begins at Home," *Nature* 564, no. 7735 (2018): S70–S71.

37. N. Caplar, S. Tacchella, and S. Birrer, "Quantitative Evaluation of Gender Bias in Astronomical Publications from Citation Counts," *Nature Astronomy* 1, no. 6 (2017): 1–5; S. Grossbard, T. Yilmazer, and L. Zhang, "The Gender Gap in Citations: Lessons from Demographic Economics Journals" (Human Capital and Economic Opportunity Working Paper 2018-078, Human Capital and Economic Opportunity Global Working Group, Chicago, October 2018), http://humcap.uchicago.edu/RePEc/hka/wpaper/Grossbard_Yilmazer_Zhang_2018_gender-gap-citations.pdf; M. A. Ferber and M. Brün, "The Gender Gap in Citations: Does It Persist?," *Feminist Economics* 17, no. 1 (2011): 151–158; G. Ghiasi, V. Larivière, and C. R. Sugimoto, "On the Compliance of Women Engineers with a Gendered Scientific System," *PLOS ONE* 10, no. 12 (2015): e0145931; M. W. Nielsen, "Gender and Citation Impact in Management Research," *Journal of Informetrics* 11, no. 4 (2017): 1213–1228; S. M. Mohammad, "Gender Gap in Natural Language Processing Research: Disparities in Authorship and Citations," arXiv preprint, arXiv:2005.00962 (2020); J. D. Dworkin et al., "The Extent and Drivers of Gender Imbalance in Neuroscience Reference Lists," *Nature Neuroscience* 23, no. 8 (2020): 918–926; A. Strumia, "Gender Issues in Fundamental Physics: A Bibliometric Analysis," *Quantitative Science Studies* 2, no. 1 (2021): 225–253; G. Abramo, D. W. Aksnes, and C. A. D'Angelo, "Gender Differences in Research Performance within and between Countries: Italy vs Norway," *Journal of Informetrics* 15, no. 2 (2021): 101114; V. Larivière et al., "Sex Differences in Research Funding, Productivity and Impact: An Analysis of Québec University Professors," *Scientometrics* 87, no. 3 (2011): 483–498; A. Paul-Hus et al., "Forty Years of Gender Disparities in Russian Science: A Historical Bibliometric Analysis," *Scientometrics* 102, no. 2 (2015): 1541–1553; M. Thelwall, "Gender Differences in Citation Impact for 27 Fields and Six English-Speaking Countries 1996–2014,"

Quantitative Science Studies 1, no. 2 (2020): 599–617; M. M. King et al., "Men Set Their Own Cites High: Gender and Self-Citation across Fields and Over Time," *Socius* 3 (2017): 2378023117738903; J. Huang et al., "Historical Comparison of Gender Inequality in Scientific Careers across Countries and Disciplines," *Proceedings of the National Academy of Sciences* 117, no. 9 (2020): 4609–4616; V. Lariviere et al., "Bibliometrics: Global Gender Disparities in Science," *Nature* 504 (2013): 211–213.

38. One of those being the analysis by Alessandro Strumia published in *Quantitative Science Studies* in 2021, which has led to an important controversy. Strumia, "Gender Issues in Fundamental Physics." See the many rejoinders published in the same issue of *Quantitative Science Studies*, "*Quantitative Science Studies*, Volume 2, Issue 1, Winter 2021: Research Articles," MIT Press Direct, accessed December 26, 2021, https://direct.mit.edu/qss/issue/2/1; as well as D. S. Chawla, "In Decision Certain to Draw Fire, Journal Will Publish Heavily Criticized Paper on Gender Differences in Physics," *Science Insider,* November 1, 2019.

39. A. Schubert and T. Braun, "Relative Indicators and Relational Charts for Comparative Assessment of Publication Output and Citation Impact," *Scientometrics* 9, no. 5–6 (1986): 281–291; L. Waltman et al., "Towards a New Crown Indicator: Some Theoretical Considerations," *Journal of Informetrics* 5, no. 1 (2011): 37–47.

40. L. Bornmann and R. Mutz, "Further Steps towards an Ideal Method of Measuring Citation Performance: The Avoidance of Citation (Ratio) Averages in Field-Normalization," *Journal of Informetrics* 5, no. 1 (2011): 228–230; Sugimoto and Larivière, *Measuring Research;* L. Waltman and M. Schreiber, "On the Calculation of Percentile-Based Bibliometric Indicators," *Journal of the American Society for Information Science and Technology* 64, no. 2 (2013): 372–379.

41. V. Larivière et al., "Team Size Matters: Collaboration and Scientific Impact since 1900," *Journal of the Association for Information Science and Technology* 66, no. 7 (2015): 1323–1332.

42. Women active in research are defined as those, based on Scopus Authors' Profiles, who contributed to at least one scientific article indexed in the Scopus database.

43. A review of counting methods can be found in M. Gauffriau, "Counting Methods Introduced into the Bibliometric Research Literature 1970–2018: A Review," *Quantitative Science Studies* 2, no. 3 (2021): 932–975.

44. See Chapter 2 for a discussion of dominant author positions. Also see, for example, P. Chatterjee and R. M. Werner, "Gender Disparity in Citations in High-Impact Journal Articles," *JAMA Network Open* 4, no. 7 (2021): e2114509; Caplar, Tacchella, and Birrer, "Quantitative Evaluation of Gender Bias"; D. W. Aksnes et al., "Are Female Researchers Less Cited? A Large-Scale Study of Norwegian Scientists," *Journal of the American Society for Information Science and Technology* 62, no. 4 (2011): 628–636; and J. R. Cole and H. Zuckerman, "The Productivity Puzzle," in *Advances in Motivation and Achievement,* ed. Martin L. Maehr and Paul R. Pintrich (Greenwich, CT: JAI, 1984), 2:217–258.

45. V. Larivière et al., "Contributorship and Division of Labor in Knowledge Production," *Social Studies of Science* 46, no. 3 (2016): 417–435.

46. Larivière et al., "Team Size Matters."

47. "Highly Cited Researchers," Clarivate, accessed December 26, 2021, https://recognition.webofscience.com/awards/highly-cited/2021/.

48. For example, C. McMilan, "Hundreds of UC Faculty Named among the World's Highly Cited Researchers," University of California, News, December 2, 2021, https://www.universityofcalifornia.edu/news/hundreds-uc-faculty-named-among-worlds-most-influential-scientists-and-scholars.

49. M. T. Nietzel, "Harvard University Employs Highest Number of the World's Most Influential Researchers," *Forbes,* December 3, 2021.

50. P. Lem, "Women Largely Absent from Asian Highly Cited Lists," *Times Higher Education,* December 1, 2021, https://www.timeshighereducation.com/news/women-largely-absent-asian-highly-cited-lists.

51. Z. Chinchilla-Rodriguez, C. R. Sugimoto, and V. Lariviere, "Follow the Leader: On the Relationship between Leadership and Scholarly Impact in International Collaborations," *PLOS ONE* 14, no. 6 (2019): e0218309.

52. Abundant evidence of this assumption is visible in the scientific literature. See L. Waltman and V. A. Traag, "Use of the Journal Impact Factor for Assessing Individual Articles: Statistically Flawed or Not?," *F1000Research* 9 (2020): 366.

53. V. Larivière and C. R. Sugimoto, "The Journal Impact Factor: A Brief History, Critique, and Discussion of Adverse Effects," in *Springer Handbook of Science and Technology Indicators,* ed. Wolfgang Glänzel et al. (New York: Springer International, 2019), 3–24.

54. M. A. Ferber, "Citations and Networking," *Gender and Society* 2, no. 1 (1988): 82–89.

55. I. Basson et al., "Gender Differences in Self-Assessment Exacerbates Inequalities in Elite Journals," in preparation (2022).

56. M. M. King et al., "Men Set Their Own Cites High: Gender and Self-Citation across Fields and Over Time," *Socius* 3 (2017): 2378023117738903.

57. G. Ghiasi, C. R. Sugimoto, and V. Larivière, "Gender Differences in Synchronous and Diachronous Self-Citations," in *Proceedings of the 21st International Conference on Science and Technology Indicators* (Valencia, Spain: Universitat Politècnica de València, 2016), 844–851, http://ocs.editorial.upv.es/index.php/STI2016/STI2016/paper/viewFile/4543/2327.

58. L. Holman and C. Morandin, "Researchers Collaborate with Same-Gendered Colleagues More Often than Expected across the Life Sciences," *PLOS ONE* 14, no. 4 (2019): e0216128.

59. Dworkin et al., "Extent and Drivers."

60. Merton, "Matthew Effect in Science."

61. P. D. Allison, J. S. Long, and T. K. Krauze, "Cumulative Advantage and Inequality in Science," *American Sociological Review* 47, no. 5 (1982): 615–625; H. Jeong, Z. Néda, and A. L. Barabási, "Measuring Preferential Attachment in Evolving Networks," *Europhysics Letters* 61, no. 4 (2003): 567.

62. M. W. Rossiter, "The Matthew Matilda Effect in Science," *Social Studies of Science* 23, no. 2 (1993): 325–341. Rossiter acknowledges that the second half of the Matthew effect was applied to describe the lack of women physicians in high positions by Lorber in her work *Women Physicians:* J. Lorber, *Women Physicians: Careers, Status, and Power* (New York: Tavistock, 1984).

63. M. W. Rossiter, "The Matthew Matilda Effect in Science," *Social Studies of Science* 23, no. 2 (1993): 327.

64. M. J. Gage, "Woman as an Inventor," *North American Review* 136, no. 318 (1883): 478.

65. Gage, 488.

66. Caplar, Tacchella, and Birrer, "Quantitative Evaluation of Gender Bias."

67. Dworkin et al., "Extent and Drivers."

68. Gage, homepage, accessed July 27, 2022, https://gage.500womenscientists.org/.

69. Cite Black Women, homepage, accessed July 27, 2022, https://www.citeblackwomencollective.org/.

70. CU Boulder School of Education Working Group for Citation Justice, "Help Us Engage in and Advance Citation Justice for Alumni and Graduates of Color," University of Colorado Boulder, School of Education, July 21, 2020, https://www.colorado.edu/education/2020/07/21/help-us-engage-and-advance-citation-justice-alumni-and-graduates-color; N. Kumar

and N. Karusala, "Braving Citational Justice in Human-Computer Interaction," *CHI EA '21: Extended Abstracts of the 2021 CHI Conference on Human Factors in Computing Systems,* May 2021, https://dl.acm.org/doi/fullHtml/10.1145/3411763.3450389.

71. P. Zurn, D. Bassett, and N. C. Rust, "The Citation Diversity Statement: A Practice of Transparency, a Way of Life," *Trends in Cognitive Sciences* 24, no. 9 (2020): 669–672.

72. J. Dworkin, P. Zurn, and D. S. Bassett, "(In)citing Action to Realize an Equitable Future," *NeuroView* 106, no. 6, (2020): 890–894.

73. D. Kwon, "The Rise of Citational Justice: How Scholars Are Making References Fairer," *Nature* 603, no. 7902 (2022): 568–571.

74. Des Jardins, *Madame Curie Complex,* 6.

7. Social Institutions

1. Men at these colleges, however, were required to be married (with some exceptions).

2. Margaret W. Rossiter, *Women Scientists in America,* vol. 1, *Struggles and Strategies to 1940* (Baltimore: Johns Hopkins University Press, 1982), 16. Barnard College was the first higher education institution in New York City to grant degrees to women.

3. L. Furumoto and E. Scarborough, "Placing Women in the History of Psychology: The First American Women Psychologists," *American Psychologist* 41 (2002): 135–142.

4. E. P. Howes, "Accepting the Universe," *Atlantic Monthly* 129 (1922): 446.

5. A. Kimball Smith, "Limited Opportunities," *New York Times,* February 6, 1983.

6. M. Baldwin, "'Where Are Your Intelligent Mothers to Come From?': Marriage and Family in the Scientific Career of Dame Kathleen Lonsdale FRS (1903–71)," *Notes and Records of the Royal Society* 63, no. 1 (2009): 86.

7. M. A. Mason and M. Goulden, "Marriage and Baby Blues: Redefining Gender Equity in the Academy," *Annals of the American Academy of Political and Social Science* 596, no. 1 (2004): 86–103; K. Michelmore and K. Musick, "Fertility Patterns of College Graduates by Field of Study, US Women Born 1960–79," *Population Studies* 68, no. 3 (2014): 359–374; E. Lazzari, "Changing Trends between Education, Childlessness and Completed Fertility: A Cohort Analysis of Australian Women Born in 1952–1971," *Journal of Population Research* 38, no. 4 (2021): 417–441.

8. M. F. Fox, "Gender, Family Characteristics, and Publication Productivity among Scientists," *Social Studies of Science* 35, no. 1 (2005): 131–150; L. Buber-Ennser and V. Skirbekk, "Researchers, Religion and Childlessness," *Journal of Biosocial Science* 48, no. 3 (2016): 391–405.

9. S. Emling, *Marie Curie and Her Daughters: The Private Lives of Science's First Family* (New York: St. Martin's, 2012), 10, 22, 56, 67, 59.

10. Emling, 27, 28, 90.

11. Emling, 89, 96, 97, 122.

12. Kathleen Lonsdale, "Women Scientists—Why So Few?," *Laboratory Equipment Digest* 86 (February 1971), quoted in Baldwin, "'Where Are Your Intelligent Mothers?,'" 87.

13. J. García Román and C. Cortina, "Family Time of Couples with Children: Shortening Gender Differences in Parenting?," *Review of Economics of the Household* 14, no. 4 (2016): 921–940.

14. T. Buchanan, A. McFarlane, and A. Das, "A Counterfactual Analysis of the Gender Gap in Parenting Time: Explained and Unexplained Variances at Different Stages of Parenting," *Journal of Comparative Family Studies* 46, no. 2 (2016): 193–219.

15. A. Saini, *Inferior: How Science Got Women Wrong—and the New Research That's Rewriting the Story* (Boston: Beacon, 2017), 5.

16. J. J. Suitor, D. Mecom, and I. S. Feld, "Gender, Household Labor, and Scholarly Productivity among University Professors," *Gender Issues* 19, no. 4 (2001): 50–67; M. Sallee, K. Ward, and L. Wolf-Wendel, "Can Anyone Have It All? Gendered Views on Parenting and Academic Careers," *Innovative Higher Education* 41, no. 3 (2016): 187–202.

17. C. R. Schwartz and H. Han, "The Reversal of the Gender Gap in Education and Trends in Marital Dissolution," *American Sociological Review* 79, no. 4 (2014): 605–629.

18. S. Cools, S. Markussen, and M. Strøm, "Children and Careers: How Family Size Affects Parents' Labor Market Outcomes in the Long Run," *Demography* 54, no. 5 (2017): 1773–1793.

19. L. A. Hunter and E. Leahey, "Parenting and Research Productivity: New Evidence and Methods," *Social Studies of Science* 40, no. 3 (2010): 433–451; S. Kyvik and M. Teigen, "Child Care, Research Collaboration, and Gender Differences in Scientific Productivity," *Science, Technology, and Human Values* 21, no. 1 (1996): 54–71.

20. J. R. Cole and H. Zuckerman, "Marriage, Motherhood and Research Performance in Science," *Scientific American* 256, no. 2 (1987): 119–125; Hunter and Leahey, "Parenting and Research Productivity"; Fox, "Gender, Family Characteristics."

21. Fox, "Gender, Family Characteristics."

22. Hunter and Leahey, "Parenting and Research Productivity"; P. L. Carr et al., "Faculty Perceptions of Gender Discrimination and Sexual Harassment in Academic Medicine," *Annals of Internal Medicine* 132, no. 11 (2000): 889–896; Kyvik and Teigen, "Child Care, Research Collaboration."

23. Hunter and Leahey examined only two disciplines (linguistics and sociology); Fox examined faculty employed in the United States in 1993–1994 across five fields; Carr and colleagues examined US medical faculty the following year (1995); and Kyvik and Teigen focused on four universities in Norway in 1992. Hunter and Leahey, "Parenting and Research Productivity"; Carr et al., "Faculty Perceptions of Gender Discrimination"; Kyvik and Teigen, "Child Care, Research Collaboration."

24. Hunter and Leahey, "Parenting and Research Productivity."

25. S. Jolly et al., "Gender Differences in Time Spent on Parenting and Domestic Responsibilities by High-Achieving Young Physician-Researchers," *Annals of Internal Medicine* 160, no. 5 (2014): 344–353; J. Misra, J. Hickes Lundquist, and A. Templer, "Gender, Work Time, and Care Responsibilities among Faculty," *Sociological Forum* 27, no. 2 (2012): 300–323.

26. Buchanan, McFarlane, and Das, "Counterfactual Analysis."

27. In collaboration with our colleagues Gemma Derrick, Pei-Ying Chen, and Thed van Leeuwen. See G. Derrick et al., "The Academic Motherload: Models of Parenting Engagement and the Effect on Academic Productivity and Performance," submitted to *Scientific Reports* (2021).

28. L. Hargens, J. C. McCann, and B. F. Reskin, "Productivity and Reproductivity: Fertility and Professional Achievement among Research Scientists," *Social Forces* 57 (1978): 154–163; Hunter and Leahey, "Parenting and Research Productivity."

29. P. Bourdieu, *Distinction: A Social Critique of the Judgement of Taste* (New York: Routledge, 1987); G. Calot and J. C. Deville, "Nuptialité et fécondité selon le milieu socioculturel," *Économie et statistique* 27, no. 1 (1971): 3–42.

30. Contemporary data still show that men with higher educational attainment have a higher number of children—except compared with those without a high school degree. V. J. Schweizer, *Fatherhood in the U.S.: Number of Children, 1987–2017*, Family Profiles FP-19-29 (Bowling Green, OH: National Center for Family and Marriage Research, 2019), https://www.bgsu.edu/content/dam/BGSU/college-of-arts-and-sciences/NCFMR/documents/FP/schweizer-fatherhood-us-no-children-fp-19-29.pdf. However, for women, it is the opposite:

the higher the educational attainment, the lower the number of children. "Women's Educational Attainment vs. Number of Children per Woman," Our World in Data, accessed July 28, 2022, https://ourworldindata.org/grapher/womens-educational-attainment-vs-fertility; G. Livingston and D. Cohn, *Record Share of New Mothers Are College Educated* (Washington, DC: Pew Research Center, 2013), https://www.pewresearch.org/social-trends/2013/05/10/record-share-of-new-mothers-are-college-educated/.

31. Derrick et al., "Academic Motherload."

32. These are respondents who indicated that they were academics and that their partner was also an academic. We did not match data between surveys—that is, we do not know whether anyone in the dataset was married to another respondent.

33. Women are also more likely to collaborate with a spouse. J. Scott Long, "Measures of Sex Differences in Scientific Productivity," *Social Forces* 71, no. 1 (1992): 159–178.

34. A. C. Morgan et al., "The Unequal Impact of Parenthood in Academia," *Science Advances* 7, no. 9 (2021): eabd1996.

35. K. A. Smith et al., "Seven Actionable Strategies for Advancing Women in Science, Engineering, and Medicine," *Cell Stem Cell* 16, no. 3 (2015): 221–224.

36. Smith et al.

37. The current expiration date for the National Institutes of Health solicitation is September 24, 2024. "Notice of Special Interest (NOSI): Primary Caregiver Technical Assistance Supplements (PCTAS) (Admin Supp Clinical Trial Optional)," National Institutes of Health, accessed July 28, 2022, https://grants.nih.gov/grants/guide/notice-files/not-ai-21-074.html.

38. Executive Committee on Research, "Claflin Distinguished Scholar Awards," Massachusetts General Hospital, accessed February 13, 2022, https://ecor.mgh.harvard.edu/Default.aspx?node_id=226.

39. R. Jagsi et al., "A Targeted Intervention for the Career Development of Women in Academic Medicine," *Archives of Internal Medicine* 167, no. 4 (2007): 343–345.

40. Smith et al., "Seven Actionable Strategies."

41. Christiane Nüsslein-Volhard-Foundation homepage, accessed February 13, 2022, https://cnv-stiftung.de/en/goals.

42. "Good News—Trainees Can Receive Childcare Support," National Institute of Allergy and Infectious Diseases, October 20, 2021, https://www.niaid.nih.gov/grants-contracts/trainees-and-childcare-support.

43. P. Vincent-Lamarre, C. R. Sugimoto, and V. Larivière, "The Decline of Women's Research Production during the Coronavirus Pandemic," Nature Index, May 19, 2020, https://www.natureindex.com/news-blog/decline-women-scientist-research-publishing-production-coronavirus-pandemic.

44. B. Lewenstein, "The Need for Feminist Approaches to Science Communication," *Journal of Science Communication* 18, no. 4 (2019): 256.

45. B. Loverock and M. M. Hart, "What a Scientist Looks Like: Portraying Gender in the Scientific Media," *FACETS* 3, no. 1 (2018): 754–763.

46. H. Mendick and M.-P. Moreau, *Monitoring the Presence and Representation of Women in SET Occupations in UK Based Online Media* (Bradford: UK Resource Centre for Women in SET, 2010).

47. D. Z. Grunspan et al., "Males Under-estimate Academic Performance of Their Female Peers in Undergraduate Biology Classrooms," *PLOS ONE* 11, no. 2 (2016): e0148405.

48. Silvia Knobloch-Westerwick, Carroll J. Glynn, and Michael Huge, "The Matilda Effect in Science Communication: An Experiment on Gender Bias in Publication Quality Perceptions and Collaboration Interest," *Science Communication* 35, no. 5 (2013): 603–625.

49. J. Terrell et al., "Gender Differences and Bias in Open Source: Pull Request Acceptance of Women versus Men," *PeerJ* preprint, e1733v2 (2016).

50. S.-J. Leslie et al., "Expectations of Brilliance Underlie Gender Distributions across Academic Disciplines," *Science* 347, no. 6219 (2015): 262–265.

51. Andrew Tsou et al., "Self-Presentation in Scholarly Profiles: Characteristics of Images and Perceptions of Professionalism and Attractiveness on Academic Social Networking Sites," *First Monday* 21, no. 4 (2016).

52. I Am a Scientist, homepage, accessed July 28, 2022, https://www.iamascientist.info/. I Am a Scientist was cocreated by Stephanie Fine Sasse and Nabiha Saklayen—the latter a physicist who has been recognized for her innovations in laser-based delivery methods.

53. Reem Alkhammash, "'It Is Time to Operate Like a Woman': A Corpus Based Study of Representation of Women in STEM Fields in Social Media," *International Journal of English Linguistics* 9, no. 5 (2019): 217.

54. A. Konnelly, "#Activism: Identity, Affiliation, and Political Discourse-Making on Twitter," *Arbutus Review* 6, no. 1 (2015): 1–16.

55. S. J. Jackson, M. Bailey, and B. Foucault Welles, "#GirlsLikeUs: Trans Advocacy and Community Building Online," *New Media and Society* 20, no. 5 (2018): 1884.

56. J. Wade and M. Zaringhalam, "Why We're Editing Women Scientists onto Wikipedia," *Nature,* August 14, 2018. Temple-Wood went on to medical school and is now a practicing physician.

57. J. Wade, "This Is Why I've Written 500 Biographies of Female Scientists on Wikipedia," *Independent,* February 11, 2019.

58. J. Adams, H. Brückner, C. Naslund, "Who Counts as a Notable Sociologist on Wikipedia? Gender, Race, and the 'Professor Test,'" *Socius* 5 (2019): 2378023118823946.

59. W. Luo, J. Adams, and H. Brueckner, "The Ladies Vanish? American Sociology and the Genealogy of Its Missing Women on Wikipedia," *Comparative Sociology* 17, no. 5 (2018): 519–556.

60. "What Do Women Do Online?," Organisation for Economic Cooperation and Development, March 2015, http://www.oecd.org/gender/data/what-do-women-do-online.htm; "Internet/Broadband Factsheet," Pew Research Center, April 7, 2021, http://www.pewinternet.org/fact-sheet/internet-broadband/.

61. M. Duggan, "Men, Women Experience and View Online Harassment Differently," Pew Research Center, July 14, 2017, http://www.pewresearch.org/fact-tank/2017/07/14/men-women-experience-and-view-online-harassment-differently/.

62. M. Kennedy, "Mary Beard Is Latest Woman to Be Sent Bomb Threat on Twitter," *Guardian,* August 4, 2013.

63. V. Metcalf, "Women Scientists Get Vocal about Top Billing on Twitter," *Conversation,* September 30, 2014, https://theconversation.com/women-scientists-get-vocal-about-top-billing-on-twitter-31906.

64. Neil Hall, "The Kardashian Index: A Measure of Discrepant Social Media Profile for Scientists," *Genome Biology* 15, no. 7 (2014): 1.

65. "By Equating Social Media Use with Narcissism, the Kardashian Index Joke Ignores Wide Disparity in Research Ecosystem," *LSE Impact Blog,* August 7, 2014, https://blogs.lse.ac.uk/impactofsocialsciences/2014/08/07/the-kardashian-index-social-media-academics/.

66. United Nations Education, Scientific and Cultural Organization, *Women in Science,* UIS Fact Sheet No. 55 (UNESCO Institute for Statistics, June 2019), http://uis.unesco.org/sites/default/files/documents/fs55-women-in-science-2019-en.pdf.

67. M. T. Brück, "Agnes Mary Clerke—Chronicler of Astronomy," *Quarterly Journal of the Royal Astronomical Society* 35 (1994): 59; M. T. Brück, *Agnes Mary Clerke and the Rise of Astrophysics* (Cambridge: Cambridge University Press, 2002).

68. Brück, "Agnes Mary Clerke," 59.

69. Brück, *Agnes Mary Clerke,* 171, 173.

70. Cited in M. Baldwin, *Making "Nature": The History of a Scientific Journal* (Chicago: University of Chicago Press, 2015), 92.

71. N. J. Silbiger and A. D. Stubler, "Unprofessional Peer Reviews Disproportionately Harm Underrepresented Groups in STEM," *PeerJ* 7 (2019): e8247.

72. D. E. Chubin and E. J. Hackett, *Peerless Science: Peer Review and U.S. Science Policy* (Albany: State University of New York Press, 1990), 87.

73. The lack of consensus in this area of research should be noted, with some research finding no difference or a difference in favor of women. F. Squazzoni et al., "Peer Review and Gender Bias: A Study on 145 Scholarly Journals," *Science Advances* 7, no. 2 (2021): p.eabd0299; C. J. Lee et al., "Bias in Peer Review," *Journal of the American Society for Information Science and Technology* 64, no. 1 (2013): 2–17; C. W. Fox and C. T. Paine, "Gender Differences in Peer Review Outcomes and Manuscript Impact at Six Journals of Ecology and Evolution," *Ecology and Evolution* 9, no. 6 (2019): 3599–3619.

74. M. Helmer et al., "Gender Bias in Scholarly Peer Review," *Elife* 6 (2017): e21718; D. Murray et al., "Author-Reviewer Homophily in Peer Review," BioRxiv, 400515 (2019).

75. Murray et al., "Author-Reviewer Homophily."

76. J. R. Gilbert, E. S. Williams, G. D. Lundberg, "Is There Gender Bias in *JAMA*'s Peer Review Process?," *Journal of the American Medical Association* 272, no. 2 (1994): 139–142.

77. R. M. Borsuk et al., "To Name or Not to Name: The Effect of Changing Author Gender on Peer Review," *BioScience* 59, no. 11 (2009): 985–989.

78. A. Tomkins, M. Zhang, and W. D. Heavlin, "Reviewer Bias in Single- versus Double-Blind Peer Review," *Proceedings of the National Academy of Sciences* 114, no. 48 (2017): 12708–12713.

79. R. M. Blank, "The Effects of Double-Blind versus Single-Blind Reviewing: Experimental Evidence from the *American Economic Review*," *American Economic Review* 81, no. 5 (1991): 1041–1067.

80. A. R. Kern-Goldberger et al., "The Impact of Double-Blind Peer Review on Gender Bias in Scientific Publishing: A Systematic Review," *American Journal of Obstetrics and Gynecology* 227, no. 1 (2022): 43–50.e4; A. E. Budden et al., "Double-Blind Review Favours Increased Representation of Female Authors," *Trends in Ecology and Evolution* 23, no. 1 (2008): 4–6.

81. "Pros and Cons of Open Peer Review," *Nature Neuroscience* 2 (1999): 197–198.

82. L. A. Rivera, "When Two Bodies Are (Not) a Problem: Gender and Relationship Status Discrimination in Academic Hiring," *American Sociological Review* 82, no. 6 (2017): 1111–1138.

83. Corinne A. Moss-Racusin et al., "Science Faculty's Subtle Gender Biases Favor Male Students," *Proceedings of the National Academy of Sciences* 109, no. 41 (2012): 16474–16479.

84. H. O. Witteman et al., "Are Gender Gaps Due to Evaluations of the Applicant or the Science? A Natural Experiment at a National Funding Agency," *Lancet* 393, no. 10171 (2019): 531–540.

85. Brück, *Agnes Mary Clerke*, 75.

86. Brück.

87. A. Saini, *Inferior*.

88. M. J. Budig and P. England, "The Wage Penalty for Motherhood," *American Sociological Review* 66, no. 2 (2001): 204–225.

89. Andrew Albanese, "ALA 2019: ALA Votes to Strip Melvil Dewey's Name from Its Top Honor," *Publishers Weekly*, June 24, 2019, https://www.publishersweekly.com/pw/by -topic/industry-news/libraries/article/80557-ala-votes-to-strip-melvil-dewey-s-name-from-its -top-honor.html.

90. Nightingale opened the Nightingale School of Nurses in 1860. It is now a faculty of King's College London under the name Florence Nightingale Faculty of Nursing and Midwifery. In fact, the enrollment of women (among other issues) led to Dewey's dismissal and the transfer of the school to Albany, where it began as part of the State University of New York. Francis L. Miksa, "The Columbia School of Library Economy, 1887–1888," *Libraries and Culture* 23, no. 3 (Summer 1988): 249–280.

91. Cassidy R. Sugimoto, Terrell G. Russell, and Sheryl Grant, "Library and Information Science Doctoral Education: The Landscape from 1930–2007," *Journal of Education for Library and Information Science* 50, no. 3 (Summer 2009): 190–202.

92. Terrell G. Russell and Cassidy R. Sugimoto, "MPACT Family Trees: Quantifying Academic Genealogy in Library and Information Science," *Journal of Education for Library and Information Science* 50, no. 4 (Fall 2009): 248–262.

93. Association for Library and Information Science Education (ALISE), *Library and Information Science Education Statistical Report 1997* (Westford, MA: ALISE, n.d.), https://ils .unc.edu/ALISE/1997/.

94. Male leadership in libraries was literally built into the profession. Carnegie Library plans in the late nineteenth century featured the men's toilet directly off the library director's office, with the assumption that the holder of this office would always be male. Abigail A. Van Slyck, *Free to All: Carnegie Libraries and American Culture, 1890–1920* (Chicago: University of Chicago Press, 1995), 15.

95. ALISE, *Statistical Report 1997*. In 2020, computer science degrees represented 25% of non–library and information science doctoral degrees among library science faculty. However, the overall composition had again shifted back to women dominated as the computer science degrees moved away from traditional library and information science programs to a growing number of iSchools coming from other orientations. ALISE, *Library and Information Science Education Statistical Report 2015*, ed. D. Albertson, K. Spetka, and K. Snow (Westford, MA: ALISE, 2015), https://www.alise.org/assets/documents/statistical_reports/2015 /alise_2015_statistical_report.pdf.

96. Nathan L. Ensmenger, *The Computer Boys Take Over: Computers, Programmers, and the Politics of Technical Expertise* (Cambridge, MA: MIT Press 2010).

97. As has been discussed in other chapters, many of these women fueled the gendered conversation about the nature of labor in these professions. For example, Grace Hopper stated that programming was "just like planning a dinner," reinforcing the gendered segregation of labor skills. Ensmenger, 73, 74.

98. Ensmenger, 77.

99. "Women in Computer Science: Getting Involved in STEM," ComputerScience.org, May 5, 2021, https://www.computerscience.org/resources/women-in-computer-science/.

100. Stuart Zweben and Betsy Bizot, "2019 Taulbee Survey: Total Undergrad CS Enrollment Rises Again, but with Fewer New Majors; Doctoral Degree Production Recovers from Last Year's Dip," *Computing Research News* 32, no. 5 (May 2020): 3–63.

101. Moshe Y. Vardi, Tim Finin, and Tom Henderson, "2001–2002 Taulbee Survey: Survey Results Show Better Balance in Supply and Demand," *Computing Research News*, March 2003, 6–13, https://cra.org/wp-content/uploads/2015/01/02.pdf.

102. Baldwin, "'Where Are Your Intelligent Mothers?'"

103. "Woman of Substance," *Chemistry World,* December 31, 2002, https://www.chemistry world.com/news/woman-of-substance-/3004326.article.

104. Dorothy Hodgkin, "Kathleen Lonsdale," *Biographical Memoirs of Fellows of the Royal Society* 21 (1976): 447–449, cited in Baldwin, "'Where Are Your Intelligent Mothers?,'" 83.

105. Baldwin, 88.

106. S. Ahmad, "Family or Future in the Academy?," *Review of Educational Research* 87, no. 1 (2017): 204–239.

107. Morgan et al., "Unequal Impact of Parenthood"; H. Antecol, K. Bedard, and J. Stearns, "Equal but Inequitable: Who Benefits from Gender-Neutral Tenure Clock Stopping Policies?," *American Economic Review* 108, no. 9 (2018): 2420–2441; L. E. Wolf-Wendel, S. Twombly, and S. Rice, "Dual-Career Couples: Keeping Them Together," *Journal of Higher Education* 71, no. 3 (2000): 291–321; J. Handelsman et al., "More Women in Science," *Science* 309, no. 5738 (2005): 1190–1191.

108. M. Oleschuk, "Gender Equity Considerations for Tenure and Promotion during COVID-19," *Canadian Review of Sociology* 57, no. 3 (2020): 502–515.

109. K. Weisshaar, "Publish and Perish? An Assessment of Gender Gaps in Promotion to Tenure in Academia," *Social Forces* 96, no. 2 (2017): 529–560; R. Durodoye et al., "Tenure and Promotion Outcomes at Four Large Land Grant Universities: Examining the Role of Gender, Race, and Academic Discipline," *Research in Higher Education* 61, no. 5 (2020): 628–651; D. K. Ginther and K. J. Hayes, "Gender Differences in Salary and Promotion in the Humanities," *American Economic Review* 89, no. 2 (1999): 397–402.

110. M. F. Fox, "Gender and Clarity of Evaluation among Academic Scientists in Research Universities," *Science, Technology, and Human Values* 40, no. 4 (2015): 487–515.

111. L. E. Wolf-Wendel and K. Ward, "Academic Life and Motherhood: Variations by Institutional Type," *Higher Education* 52, no. 3 (2006): 487–521; Ahmad, "Family or Future?"

112. Ahmad, "Family or Future?"

113. S. Wilton and L. Ross, "Flexibility, Sacrifice and Insecurity: A Canadian Study Assessing the Challenges of Balancing Work and Family in Academia," *Journal of Feminist Family Therapy* 29, no. 1–2 (2017): 66–87.

8. Recommendations and Conclusions

1. P. Zurn, "Feminist Curiosity," *Philosophy Compass* 16, no. 9 (2021): e12761.

2. V. Larivière et al., "Sex Differences in Research Funding, Productivity and Impact: An Analysis of Québec University Professors," *Scientometrics* 87, no. 3 (2011): 495.

3. J. Dworkin, P. Zurn, and D. S. Bassett, "(In)citing Action to Realize an Equitable Future," *Neuron* 106, no. 6 (2020): 890–894.

4. D. Kozlowski et al., "Intersectional Inequalities in Science," *Proceedings of the National Academy of Sciences* 119, no. 2 (2022): e2113067119.

5. C. Rovira, L. Codina, and C. Lopezosa, "Language Bias in the Google Scholar Ranking Algorithm," *Future Internet* 13, no. 2 (2021): 31.

6. Women in Neuroscience homepage, accessed February 13, 2022, https://www.winrepo .org. There were 1,864 entries as of February 13, 2021.

7. Anneslist homepage, accessed February 13, 2022, https://anneslist.net/; "Women in Cell Biology," American Society for Cell Biology, accessed February 13, 2022, https://www .ascb.org/committee/ascb-women-in-cell-biology-committee/.

8. C. R. Sugimoto and V. Larivière, "Indicators for Social Good," Centre for Science and Technology Studies, blog archive, May 15, 2019, https://www.cwts.nl/blog?article=n-r2w2c4; European Commission, Directorate-General for Research and Innovation, *Indicators for Promoting and Monitoring Responsible Research and Innovation: Report from the Expert Group on Policy Indicators for Responsible Research and Innovation* (Luxembourg: Publications Office of the European Union, 2015), https://data.europa.eu/doi/10.2777/9742.

9. C. R. Sugimoto and V. Larivière, *Measuring Research: What Everyone Needs to Know* (New York: Oxford University Press, 2018).

10. T. Obloj and T. Zenger, "The Influence of Pay Transparency on (Gender) Inequity, Inequality, and the Performance Basis of Pay," *Nature Human Behavior* 6 (2022): 646–655.

11. L. L. Taylor et al., "How to Do a Salary Equity Study: With an Illustrative Example from Higher Education," *Public Personnel Management* 49, no. 1 (2020): 57–82.

12. Known as Goodhart's law: "Any observed statistical regularity will tend to collapse once pressure is placed upon it for control purposes." C. A. Goodhart, "Problems of Monetary Management: The UK Experience," in *Monetary Theory and Practice* (London: Palgrave, 1984), 96.

13. P. Wouters et al., "Rethinking Impact Factors: Better Ways to Judge a Journal," *Nature* 569 (2019): 621–623.

14. L. Zhang et al., "Should Open Access Lead to Closed Research? The Trends towards Paying to Perform Research," *Scientometrics* (2022): 1–27.

15. J. Schroeder et al., "Fewer Invited Talks by Women in Evolutionary Biology Symposia," *Journal of Evolutionary Biology* 26, no. 9 (2013): 2063–2069.

16. A. Casadevall and J. Handelsman, "The Presence of Female Conveners Correlates with a Higher Proportion of Female Speakers at Scientific Symposia," *MBio* 5, no. 1 (2014): e00846-13.

17. E. Edwards, "No More 'Manels,' NIH Head Says in Call to End All-Male Science Panels," NBC News, June 13, 2019, https://www.nbcnews.com/health/health-news/no-more-manels-nih-head-says-call-end-all-male-n1017181.

18. "About," BiasWatchNeuro, accessed February 13, 2022, https://biaswatchneuro.com/about/.

19. H. Else, "How to Banish Manels and Manferences from Scientific Meetings," *Nature* 573, no. 7773 (2019): 184.

20. C. Russell, "Women Science Writers Conference about Changing the Ratio," *Colombia Journalism Review,* June 18, 2014, https://archives.cjr.org/the_observatory/science_writing_rebranding_fem.php.

Appendix

1. C. R. Sugimoto and V. Larivière, *Measuring Research: What Everyone Needs to Know* (New York: Oxford University Press, 2018).

2. P. Mongeon and A. Paul-Hus, "The Journal Coverage of Web of Science and Scopus: A Comparative Analysis," *Scientometrics* 106, no. 1 (2016): 213–228; V. Larivière et al., "The Place of Serials in Referencing Practices: Comparing Natural Sciences and Engineering with Social Sciences and Humanities," *Journal of the American Society for Information Science and Technology* 57, no. 8 (2006): 997–1004.

3. M. Lacroix and V. Larivière, "Parité de genre et genre de parité dans les revues québécoises," in *Relire les revues québécoises: Histoire, formes et pratiques,* ed. É. Guay and R. Nadon (Montreal: Presses de l'Université de Montréal, 2021), 305–317.

4. K. Hamilton, *Subfield and Level Classification of Journals,* no. 2012-R (Cherry Hill: CHI Research, 2003).

5. "Standard Country or Area Codes for Statistical Use (M49)," United Nations, accessed May 2021, https://unstats.un.org/unsd/methodology/m49/.

6. V. Larivière et al., "Bibliometrics: Global Gender Disparities in Science," *Nature* 504, no. 7479 (2013): 211–213.

7. See V. Larivière et al., "Supplementary Information To: Global Gender Disparities in Science," *Nature,* December 12, 2013, https://static-content.springer.com/esm/art%3A10.1038%2F504211a/MediaObjects/41586_2013_BF504211a_MOESM105_ESM.pdf.

8. See Larivière et al., "Supplementary Information," table S6.

9. "Gender Statistics," World Bank, accessed December 15, 2020, https://datacatalog.worldbank.org/search/dataset/0037654.

10. "Gender Data Portal," World Bank, accessed December 15, 2020, https://www.worldbank.org/en/data/datatopics/gender.

11. J. P. Birnholtz, "What Does It Mean to Be an Author? The Intersection of Credit, Contribution, and Collaboration in Science," *Journal of the American Society for Information Science and Technology* 57, no. 13 (2006): 1758–1770.

12. A different editorial system for *PLOS Biology* made it difficult for PLOS to provide us with the data for this journal. Therefore, while the *PLOS Biology* contributorship data are included in the global analysis, individual data for the journal are not provided.

13. Larivière et al., "Bibliometrics."

14. V. Larivière et al., "Contributorship and Division of Labor in Knowledge Production," *Social Studies of Science* 46, no. 3 (2016): 417–435.

15. "Download Awards by Year," National Science Foundation, accessed November 10, 2021, https://www.nsf.gov/awardsearch/download.jsp.

16. P. Zurn, D. Bassett, and N. C. Rust, "The Citation Diversity Statement: A Practice of Transparency, a Way of Life," *Trends in Cognitive Sciences* 24, no. 9 (2020): 669–672.

ACKNOWLEDGMENTS

This book was a labor of love and would not have been possible without the support and encouragement of friends, family, and colleagues around the world. First and foremost, we would like to thank Faith Bosworth for propelling the project forward through every barrier. This book simply would not exist without her. Thanks to Janice Audet, our editor at Harvard University Press, who believed in the project from the start and never gave up on us. Gratitude also goes to Gemma Derrick and Kalpana Shankar, who convinced us that there was an audience for this work, and to Amanda Morgan for her encouraging words.

Thanks also to the team at the Observatoire des sciences et des technologies (Pascal Lemelin, Benoît Macaluso, Jean-Pierre Robitaille, and Mario Rouette) who maintained the bibliometric database, and to Philippe Vincent-Lamarre for data structuration. Thanks also go to our research teams, who have provided feedback along the way, and particularly Michaela Krawczyk and Yong In Choi, who helped with reference formatting. Gratitude goes to Gemma Derrick (again), Mary Frank Fox, Jean A. Pratt, and Seokkyun Woo for feedback on early drafts of the manuscript. A big thanks to Christine Davis, whose keen eye helped with the final polish and formatting, as well as to the reviewers, whose comments helped to strengthen the book.

We would like to thank all the institutions that provided us time or space to work on this project, with a special thanks to Indiana University Bloomington, Université de Montréal, the National Science Foundation, Université du Québec à Montréal, Georgia Institute of Technology, Leiden University, and Stellenbosch University. This book is truly an international collaboration, written at these institutions, on long-haul flights, and in hundreds of coffee shops and airport terminals around the world. However, our greatest sites of productivity were in Bloomington, Indiana; Atlanta, Georgia; Leiden; Montréal; McLean, Virginia; Aucun, France; and Portland,

Maine. The first full draft of the book was concluded in Stellenbosch, South Africa, after two years of various levels of pandemic-imposed quarantines.

Finally, we owe an enormous debt of gratitude to all our collaborators for the past decade. These authors are cited where the relevant work is mentioned, but the references do not fully account for all the tacit knowledge that has been exchanged through each of these interactions. We are grateful for the opportunity to have worked with each of these scholars. This is especially true for our students, who teach us more than they may ever know and bring an incredible richness to our professional lives.

INDEX

Note: The letter *t* following a page number denotes a table. The letter *f* following a page number denotes a figure. The *n* following a page number denotes a Note followed by the note number.